The Story Economy

The Story Economy

HOW SHARING YOUR TRUTH CAN BE YOUR MOST VALUABLE ASSET

William Welser IV

RADIUS BOOK GROUP
NEW YORK

Radius Book Group
A Division of Diversion Publishing Corp.
www.RadiusBookGroup.com

Copyright © 2025 by William Welser IV

All rights reserved, including the right to reproduce this book or portions thereof in any form whatsoever. No part of this publication may be reproduced or transmitted in any form or by any means, electronic or mechanical, including photocopying, recording, or any other information storage and retrieval, without the written permission of the author.

For more information, email info@radiusbookgroup.com.

First edition: October 2025

Trade Paperback ISBN: 9798895150801

eBook ISBN: 9798895150818

Manufactured in the United States of America

1 3 5 7 9 10 8 6 4 2

Cover design by Will D. Mack

Interior design by Neuwirth & Associates, Inc.

Radius Book Group and the Radius Book Group colophon are registered trademarks of Radius Book Group, a Division of Diversion Media and Communications, LLC

Dedicated to Onyx of the Purple Papi Sox. Not just man's best friend but my copilot for every late night, all nights. Please excuse any lapses in the text—those moments were 3 a.m. drive-by pawings when he nudged my laptop closed and herded me to bed.

CONTENTS

A TINY DISCLAIMER • IX

PART 1 • Why We Need Saving

INTRODUCTION: Can Stories Save Us? • 3

CHAPTER 1: A Disclaimer on Duality • 15

CHAPTER 2: We're in Trouble • 22

CHAPTER 3: We're Lonely • 31

CHAPTER 4: We're Broke • 43

CHAPTER 5: We're Lost • 53

CHAPTER 6: We're Losing Control • 67

CHAPTER 7: We're Not Even Real • 87

CHAPTER 8: We're on Our Own • 113

PART 2 • The Power of a Single Story

CHAPTER 9: Letting It All Out • 123

CHAPTER 10: The Stories Inside • 136

CHAPTER 11: Cognitive Bias • 145

CHAPTER 12: Unconscious Bias • 154

PART 3 • The Story Economy

CHAPTER 13: The Idea • 163

CHAPTER 14: The Possibilities • 170

CHAPTER 15: The Tech • 179

CHAPTER 16: The Currency • 184

CHAPTER 17: The Marketplace • 190

CHAPTER 18: The Context • 197

CHAPTER 19: The Insights • 204

CHAPTER 20: The Ledger • 209

CHAPTER 21: The Future • 213

CHAPTER 22: The Tasks • 221

APPENDIX: *Pick Your Misperception* • 241

NOTES • 251

ACKNOWLEDGMENTS • 261

A TINY DISCLAIMER

This book is outdated.

It was already antiquated the minute I typed these words onto these pages.

This book covers current advancements in technology, and the speed at which those advancements are happening is faster than any other type of growth we've ever seen in any other industry. In the time it took to write this very small paragraph, an AI somewhere has taken in another hundred million data points and is making decisions with that new information.

I say this not to scare you but to help you. I want you to take your power back from Big Tech, and knowledge is the best way to do so. I want you to understand what's going on with you and your data, and what you can do about it. Unfortunately, advancements in regulating the practice of stealing your privacy aren't gaining as much traction. I hope this book gives you the power to gain more agency over your data, your devices, your decisions, and your life.

PART 1

Why We Need Saving

INTRODUCTION

Can Stories Save Us?

2016—Venice Beach

I was sitting with some really smart colleagues from RAND—a nonprofit think tank—outside of an Erewhon supermarket in Venice Beach. If you want to understand the wealth gap, this uber posh grocery store is a great place to start. You have to pass several homeless encampments to get to the automatic glass door, yet when you step inside, it feels almost necessary and cool to spend $30 on an ounce of almonds. I don't because I'm allergic, but I digress.

We were sitting outside having coffee (not almonds). Our regular lunchtime ritual was trying to solve the problems of the world. There was no better spot for this discussion than a well-branded "luxury wellness" supermarket in which customers are led through like maze rats searching for that last wheel of plant-based, cave-aged truffle brie and other obscure goodies that celebrities post about on Instagram.

We didn't look as cool as the typical Erewhon shoppers. None of us had an Instagram account at the time, nor had any of us ever tried on a piece of Lululemon clothing. We wore smart basics with pockets and comfortable shoes. Feel-good science geeks, we were experts on obscure subject matters like empathy, neural pathways, habit forming, and the effects of technology on humanity.

We worked at RAND and in academia in an attempt to help inform public policy and use our intelligence to do something important and change the world. As a principal investigator, I was fortunate enough to be put in charge of a team of extremely talented researchers, engineers, and scientists bent on healing the world.

It was a volatile time for our country. Bernie Sanders, that mitten-wearing, outspoken progressive, was among the candidates vying to gain the Democratic Party win in the presidential primaries and run against Donald Trump. He'd made the redistribution of wealth, which he called a disgrace, a central component of his campaign. He referred to the wealth gap as the great moral issue of our time.

We agreed with his platform and respected the depth of his perception and his down-home attire. We appreciated any attempt to identify some of our biggest problems.

"Just look around us," said Emily, a feisty brunette with a PhD in psychology and neuroscience. "How is it fair to live in a society in which some people pay thirty dollars for almonds and some can't pay their electric bill?"

I looked at the *Erewhon* sign above the doors where fiercely trendy customers entered the market. It was an anagram for Nowhere, an allusion to a satirical 1872 novel about health written by Samuel Butler. You didn't have to look far to find the modern-day satire in Erewhon the grocery store—a place that went viral in 2025 for selling one single Japanese strawberry in a cupcake container for $19.99.

We read the packaging strewn about our table. There were so many buzzwords: Regenerative. Non-GMO. Organic.

"The fact that you have to be wealthy to afford to eat food that isn't a serious health hazard feels unjust," offered Brett, another PhD and an expert in machine learning and cryptography. "Being broke isn't good for our constitution or our morale."

I looked past a table of three thin, ultra-trendy blondes taking selfies, their skin glowing probably from the bee pollen in their $20 smoothies. In plain sight was a tent city, rows of rough encampments of unhoused people created out of tarps, tents, and all manner of salvaged objects. I imagined that the sight of wealthy, pretentious people walking their carts through the grocery store could have a particularly demoralizing effect on a guy collecting cans in his cart or the mother living on the street who couldn't afford to feed her kids.

This is the world we live in. It is inhumane. How is it fair that some people can have so much while so many must go without? I'd venture to guess Bernie Sanders himself wouldn't be able to handle such a juxtaposition.

As usual, our conversation quickly progressed after we'd identified the problems. We talked about how we knew the politicians fighting about their strategies weren't going to solve it. Sanders was envisioning a Robin Hood type approach, where he'd tax the rich and redistribute the excess to people with less. Donald Trump proposed the opposite, giving more tax breaks to the wealthy, and I'm still not quite sure what Hillary Clinton was planning to do to help the "deplorables." Either solution would create a winner and a loser. Take from the rich to give to the poor, and the rich feel like the losers. Give tax breaks to the rich, further the socioeconomic divide, and the poor feel like the losers.

Any time you take money from any side, you're making someone say, "Hey! That's not fair," even if the people you're taking from have so much money they'd never be able to spend it anyway. In my experience, many billionaires will do whatever it takes to save a buck. Redistributing their many bucks, and forcing them to do it, would not be done without a fight.

Sitting outside the most expensive grocery store in America and watching the wealthiest people walk in and out of the glass door, we decided that our objective must be to find a solution in which everyone could come out feeling like a winner. We needed a neutral-win situation or a win-win situation. Nobody could feel like the loser. We used our collective intelligence and specialties to come to the conclusion that the only way we could see a win-win outcome would be to come up with a different form of currency.

What if we gave everyone the ability to purchase something without having to use money? The new form of currency would have to be something that either everyone could get easily or something that everyone already had. What does everyone have that could be worth something? The adage states that we only have death and taxes in common, and neither of those sounded any fun to monetize. We all have skin and organs, but putting a value on those seemed a little barbaric. We ultimately realized that the most valuable thing that every human possesses is something special and unique to each one of us. When used correctly, it could be much more valuable than dollars. We all have millions of them brewing inside our heads and filling our hearts.

Stories.

We all have the beautiful and harrowing experiences that have made us into who we are. Each little story within those experiences could be extremely valuable to ourselves and to corporations. Our stories can create data, and everyone knows data is the hottest commodity in the world right now.

Stories could be the new currency. Stories could help us buy things.

And what perfect timing, because doesn't it seem we could use an alternative economy right now?

I don't know about yours, but my current LinkedIn feed is a sea of out-of-work, pre-retirement professionals wondering if they should learn carpentry. The future of many industries seems to be hanging by the thinnest of threads. Bill Gates himself has said that in the future, "Humans won't be needed for most things." My visit to Erewhon was in 2016, and with the technological advancements over even just the past few years, there's a higher chance most of us end up in the homeless encampment instead of buying $20 strawberries. If there's any time for an alternate economy, I'd say it's now. No, I'd yell it loudly: WE NEEDED A NEW KIND OF ECONOMY YESTERDAY!

How about a story economy? Does this sound crazy? The idea of buying something with a story? Maybe it is! My colleagues and I thought it could really change everything we understood about the structure of our very society.

After lunch that day, we all went our separate ways, ready to solve another global challenge at lunchtime the next day. I couldn't stop thinking about using our stories as currency. How different the world could be if everyone had the same opportunity to build wealth and/or to be treated the same upon walking into a store, whether your bank account has $10 or $10 million. It would change how we buy things. It would change how we see each other. It would rearrange capitalism. It would solve so many problems and help so many people.

I didn't sleep that night, wondering if a story economy could be possible.

THE NEXT MORNING, I filled a notebook with ideas. I highly recommend the Leuchtturm1917 hardcover notebooks with dotted pages. (My entire career sits in those on my office bookshelf.)

The idea of using stories as currency is the only idea out of the kajillions we'd come up with over the years that felt like it could satisfy both billionaires and the rest of us. There's an unlimited supply of stories, and brands would pay top dollar for them.

I spent the day imagining all the possibilities.

First, I had to remind myself why stories were so important. I wrote about ideas I'd gleaned from the thousands of people I'd interviewed at RAND, an organization devoted to creating safer and healthier communities through public policy.

While it is true that many of our problems are epitomized by the discordance between the virtual cart-filling millionaires and the physical cart-pushing street campers—there are deeper issues underlying that polarity.

The gaping divide between the rich and poor is no longer definable through brick and mortar. Billionaires aren't limited to classic chess moves like buying buildings or stowing their money in offshore accounts. Once upon a time our lives would not be directly impacted by the richest among us. Sure, we may have had to work a dead-end office job to keep our health insurance. Yet, we had more autonomy. We could choose to refrain from investing in the stock market or buying a high-end pair of sneakers we couldn't afford. Now the richest people in our country are the billionaires and we are subject to technological psychological bombardment from their companies and the advertisers, influencers, and white-collar scam artists who piggyback on their campaigns and data. It happens every time we get lost mindlessly scrolling on our devices (which, unless you're a member of an undiscovered tribe in the Amazon, is likely quite a few hours every day).

This impacts us in alarming ways.

I thought back to those girls with the glowing skin and expensive Erewhon smoothies snapping their selfies. There was an inauthenticity about them. I thought about the illusion that drew people to buy $20 drinks and cave-aged not-cheese; to ignore the discordance between the hyper-healthy obscenely overpriced food and the people who were starving around them. And I thought of my own inauthenticity occasionally, the way I often blindly repeat answers to questions by rote—without any kind of emotional connection. I thought about all the important conversations I've had, where my mind and eyes were inadvertently drawn to my screens instead of my three children, Alaina (twenty), William (eighteen), and Calvin (sixteen), and my wife, Eileen.

TODAY'S WORLD ACTUALLY sets us up to look away from our lives. We spend hours and hours a day scrolling our phones looking at *other* people's lives instead. In the process, we are looped into a vortex of social media and internet advertising, livestreams and clickbait deliberately designed to dysregulate and distract us. And as we scroll, we click, we buy, we compare ourselves . . . and after all that, we don't feel very good. Big Tech—the technology giants like Meta, which runs Instagram and Facebook, Apple, Twitter (fine . . . X), TikTok, and any other tech company vying for our attention—wants it that way. They do whatever they can to trick us into returning to our devices, perpetually, habitually, so we continue to scroll, click, buy, and compare ourselves more and more.

Avoiding our true stories is hurting us. It's breaking us. It's actually killing us.

Over the years, most of us have fallen into the trap of spending more time online and less time in real life. The average Gen Z phone user is on their screen for around nine hours a day. Most of them search TikTok first when they want to know something I would have looked up in my dad's encyclopedia when I was that age. And almost half of Gen Z watches Netflix (or some other streaming media) on their phones over four hours a day![1]

When social media first hit the scene, it felt great. We found old friends, discovered a new way to find out if someone was single, bragged about our delicious dinners, and shared memes that made people laugh. But as technology grabbed hold of us, its importance in our everyday lives shifted. It was no longer intended to be fun or to give us the opportunity to take a quick break from work. It became habitual. Our screen time is soul sucking, something we can't turn away from. In many ways, we have lost autonomy over our own lives.

In the eyes of many tech companies, we are wallets, not humans. Those Big Tech giants like Instagram, Facebook, TikTok, and X have done a great job designing their apps and devices to keep us stuck to them like mice to glue traps. They love when we click more and spend more. In fact, with the collection of our data and the monetizing of our clicks and scrolls, we are no longer using technology. It's using us.

I WANT YOU to be aware of what's happening. I want you to realize that you (and me and most of humanity) are (figuratively speaking) stuck in a wildly treacherous body of rushing water, and it's scary. Big technology companies opened those floodgates and are inviting us aboard their huge cruise ships offering to navigate those waters for us—the very rapids that they create. We are allowing them to do that.

What other options do we have? Give up? Drown? Stand on the shore while the world rushes by?

This loss of power has created three main issues plaguing society today:

We are lost.

We are lonely.

We are broke.

It's difficult to experience just one of those issues, and many of us are constantly dancing with all three. We either don't know ourselves or we are perpetually questioning ourselves. We don't feel like we can turn to a supportive community for help, and we can barely think about any of it because we're working so hard to make ends meet. It's painful. It's confusing. And it has to end.

The conversation with my colleagues made me believe, for the first time in a long time, that there was a way to actually empower ourselves while still using technology.

Changing our relationship with technology can spur us to be brave, informed navigators through those rushing waters. Instead of being swept away, we can learn how to navigate technology correctly using tools and perspectives that can change everything. We can learn to take our power back and direct it toward exercising agency over our interests.

I've spent so much time pondering this and experimenting with how we can make it work. Finally, after almost ten years, I feel like I might have found a solution. And, indeed, it starts with a story.

Stories are our source of great connection.

They always have been.

THE ORIGIN OF STORY

The *Epic of Gilgamesh* is potentially the oldest written story, and it still helps readers ponder their existence and question the world all these years later. Originating in Mesopotamia in 2100 BCE and rediscovered in the nineteenth century during the excavation of an ancient king's library, it follows the adventures of Gilgamesh, a powerful yet arrogant king who was two-thirds god and one-third human.

Gilgamesh oppressed his people by forcing them to work in unsafe conditions, taking women as property, and not caring about or understanding the needs and feelings of others. The gods didn't like this. They responded by creating Enkidu, a wild man of the steppe, to challenge him. After a fierce wrestling match, they became close friends and embarked on heroic adventures together.

Unfortunately, Enkidu fell ill and died. His death awakened Gilgamesh's fear of mortality. Gilgamesh journeyed to find eternal life but ultimately learned that immortality is reserved for the gods and that he must accept his humanity and mortality. That is sure relatable to me, a man in his forties who is beginning to come to terms with his own mortality more than four thousand years later!

This story connects me to people I will never meet (both the characters and the author) and connects me to the past. It helps me to imagine ancient people who were also wrestling with thoughts about the human condition, mortality, and the search for meaning. Someone chose to write this tale down in cuneiform on tablets, because they thought it was important to record.

This one powerful story demonstrates that humans have been on a quest to find the same things for thousands of years. We want intimacy and friendship. We want to matter. We want to be remembered. We want to be the opposite of lost, lonely, and broke. We always have.

The valuable connection and meaning found in almost any true story are the reason I have become a story collector. I record stories. I listen to stories. I ask people stories. I live for stories. Stories hold so much context that they can change everything. Various research and technology projects at RAND involving stories have taught me what people need, how they feel, and what we can do about those needs and feelings as we build technology around their experiences. When we hear an entire story rather than a simple fact, we learn so much and have so much more available to us to make better decisions or to understand people and the world at large.

When my daughter, Alaina, was about to graduate high school, she read the beautiful pamphlet about her prospective university and thought it looked nice. The smiling photos made her curious, for sure. She was heartened by the facts listed in bold: 83.9 percent of students graduate after four years. She imagined herself starring in the photos of the various departments: debating in the auditorium, performing onstage, wearing lab glasses and hunched over a beaker.

Or rather, she wanted to imagine it. How would she know if her actual experience could resemble the experiences of any of the students in the photos?

What was it *really* going to be like for her?

She couldn't quite tell. But as I drove her back from an overnight visit to see the school in person, she was buzzing with excitement, telling me about all the stories the current students had told her. She couldn't wait to explore the city after classes and camp out on the lawn for special events. Before that visit, the stories on the pamphlet had her pretty sure she wanted to attend that university. After she heard other students' stories in person, she was ready to enroll. She signed up immediately, and I drove her (stifling my tears) to her dorm not too long afterward.

The details I mentioned in this story about me and my daughter are important. I chose it to connect with you. I want you to know more about me, William—the guy on the other side of these pages. I want to connect with you, so I added a bit more context. It's a fact that I drove my daughter to her dorm. But the context for this story was that I was holding back tears as I did it, that I was excited for my daughter but grieving the loss of the child she once was.

That tiny bit of extra info allows you to get to know me a little better. You can at least guess that I'm a softie. You may recognize that at the very least I love my kids, and I wasn't just chopping onions in the kitchen before we hopped in the car to go to college.

What else can I tell you about me? I'm William Welser, the Fourth. Yes, the Fourth. While it sounds like an aristocratic title, I actually come from a huge military family and a long line of men and women committed to protecting and helping other humans. While I may have chosen to cut my time as an officer in the Air Force shorter than the rest of my family members (sorry, Dad), I have still devoted my life to helping and protecting. I spent a decade as a principal investigator, professor, and executive at RAND. I've executed all kinds of experiments to study how technology affects humans and how to change people's lives for the better. And now I am running a tech company called Lotic, hoping to create a technology that will tap into stories to shift how we engage with ourselves, corporations, and social media. It's another way to protect and serve.

But, most importantly, I'm a dad. I want my children to be able to thrive in the future. I have seen some concerning things through my research. This book is my humble attempt to add to the dialogue that will allow us to make the societal changes we need to thrive as a society. I'm an insider to the AI/tech research world waving my hands and asking you to pay attention. And to tell more stories.

CAVE INFLUENCES

Stories help us make decisions, help us understand who we are, and help us leave a legacy. We've been wanting that for so long. In 2020, a team of scientists found what they believe is the oldest cave painting discovered to date—dating back to forty thousand years before Gilgamesh. It depicts a figurative drawing of a Sulawesi warty pig—a wild boar endemic to

South Sulawesi. Nearby, they discovered an entire scene depicting hybrid human-animal beings hunting Sulawesi warty pigs. What was the story they were wanting to tell? Were the people of that community hoping to make a point about their hunting prowess? Or were they simply telling a story about their day? Was this an ancient Instagram and this was their way of influencing? Or were they warning of the dangers of the day? Maybe it was an advertisement or a simple note. From the tale they told on those cave walls during a time I can only imagine, I can still connect with them, forty-five thousand years later. They hunted pigs. They were proud of themselves. They wanted someone to know about it.

I can relate. I am sitting down now typing out the story of my own life, my own cave painting, to attempt to communicate something to you about the way I make meaning in my own life. My own story may at times seem crude, cryptic, or unenlightened. It may illustrate my own prowess or cowardice. It is my own SOS, my warning to avoid the dangers of our day. It is a story I began carving out in 2016 when my copious data-collecting and research led to the realization that so many of us are struggling with feelings of loneliness, estrangement from community, and lack of financial freedom.

In that sense my story is your story, the story of what is happening in our world. I believe that we are at a crossroads, that we have a critical choice to make. We can keep going down the dangerous path we are on, blindly bait clicking and screen shaming, led like Pavlov's dogs toward distractions and dysfunction. All the while our information is harvested by corporations whose only motivation is to line their own pockets and who do not have our own best interests at heart (not by a long shot).

We can keep comparing our lives (and our faces and bodies) to the imaginary lives Instagram influencers have led us to believe we should covet. Or we can be brave and learn what is happening to us, and shake off the technological spell that has been cast. We can learn a new way of existing alongside technology. We can empower ourselves with it. We can use it in a new way in which we have better control.

Over the course of our time together within this book, I hope I'll relay to you how we got so derailed and how we fell out of sync with our true natures and became lonely, broke, and out-of-control. Most important,

I hope I can convince you that you have the opportunity to shift this dynamic through the power of your own story. Stories hold enough power to pull us out of our personal abyss. We can use them to empower ourselves and prepare for a big societal shift. We have the ability to make big changes in our lives so we can all walk into a future with agency, community, and more money in our wallets.

I'm convinced . . .

Stories can save us all.

CHAPTER 1

A Disclaimer on Duality

I spent a weekend making this little bench for my plants. I live alone in San Francisco at the top of a steep hill. People often walk up the hill and stare at my house. I made this little bench to give some of my potted plants a lift so they would cover my window, and all the tourists wouldn't look at me. I'm not in my undies all the time or anything, but I like my privacy. I would say I'm an INtrovert, so I want to be "IN" my house and left alone. Anyway, I took some other wooden pots and part of a wooden bookshelf that had been deteriorating in my garage, and I made a bench. I went inside for just a bit so I could relax before lifting the heavy plant pots up, and bam, two tourists were sitting on my damn bench. I was like, what? How? Wait? But then I figured maybe they really needed to rest. When I finally came outside again to put the plants up, there were more tourists resting on my bench. I decided to watch and see over the next few days, and my bench has now become a neighborhood watercooler. I mean, people meet there. People rest there. Dogs sniff around. It's just a few pieces of old wood, but I guess we needed it. People needed it. I have since put some actual nails in it so it wouldn't break under the weight of all the neighbors. And I painted it. It looks more inviting, even though I'm not actually inviting anyone to talk to me or look at me. I'll still remain safe inside. But everyone else can have my bench.

—Anonymous, audio-recorded story

I was in my living room joking with my sixteen-year-old son, Calvin, about a funny dunk shot his teammate made in their last basketball game. The phone rang, abruptly. I picked it up off the end table and looked at my screen. *Dammit,* I thought seeing that it was Rick, an employee I'd hired who had been performing poorly. Rick was responding to a message I left chastising him for being absent and not responding to his DMs.

I clicked on the phone.

"Rick, what the hell?" I said harshly.

I looked at my son and we both laughed silently at how quickly I changed my tone. I became an evil boss in an instant.

We're going to tackle a lot of issues in this book, and the first is the fact that I can be kind of a jerk. But I am also a very nice guy. And guess what? You can also be kind of a jerk and kind of a nice person. So can everyone. We can't help it. We all have duality. Everything in our society has more than one side to it. That is one reason we've found ourselves in a bit of a tumultuous time.

This is also one reason I love humanity and want the best for us all. The ability to be multi-faceted or to always recognize another perspective is part of our magic.

I used to love taking my kids to parks when they were little. It was always such a surprise to see what they would do and how they would find another use for something meant to be used in another way. I wanted the garbage can to be used for garbage. Well, when my son William (yes, William the Fifth) was five, he thought it was best turned over and used as a stage for his rendition of *Peter Pan.* My wife, Eileen, wanted the kids to swing on the swings. Nope. My daughter, Alaina, and son Calvin decided they were best used as little desks for playing office. Kids are so good at reinvention. They have that wild creativity that is sometimes quashed in us once we arrive at adulthood. Is it a leaf, or is it a spaceship for a bug on its way to the moon? The ability to see an object or idea from several points of view is a super talent. And it's marvelous that our brains can do this.

However, creation and innovation have their downfalls.

As a technologist in the midst of creating a brand-new kind of tech based on my years and years of researching humans and their use of technology, I am still quite shocked that most technologists (including me) tend to be super optimistic about their creations. They think that people will use their app or software exactly how they'd imagined it would be used. They rarely consider the bad outcomes. They don't recognize the duality of everything, and instead just assume the best. This belief is adorable. It can also be harmful. We must always assume that someone will figure out another use for our creations—malicious, nefarious, or otherwise. It's also important to work hard to predict what those uses might be.

Who could have predicted what's happened to platforms like Instagram? Were you around when it first debuted? Suddenly there was this brightly colored app on your fairly new smartphone that let you share pretty photos with your friends and family. You could even get creative with it and make your pictures look vintage through a cool filter. Amazing. It seemed like a total upgrade from the MySpace everyone was using that required so much reading. This would be purely visual and absolutely captivating.

Unfortunately, the other side of showing off photos is just showing off. An entire industry sprouted up to cater to our penchant for exhibitionism. Platforms like Instagram (among its counterparts like Facebook and those that came before . . . maybe even Friendster) sparked influencers and cyberbullying and body shaming and steered followers toward anorexia, suicidal ideation, dopamine addiction, misinformation, and more and more content that keeps us from sleeping well at night. We sure didn't consider the *other* futures made possible by photo sharing apps.

Of course, it's mostly impossible to predict the future—so impossible that I try to avoid using the "p-word" and say "anticipate" instead. Kevin Systrom and Mike Krieger, the creators of Instagram, may have thumbtacked an index card with the phrase "get bought by Zuckerberg" on their vision board, but they had no idea their photo sharing app would become such an influencing behemoth. The creators of spray paint had no idea it would be used to ruin, embellish, tag, or create underground art on everything from subway tunnels to fifty-foot billboards all over the world. The scribes of Catholicism had no idea priests, leaders, and other officials would sometimes use that religion to shame people and make them feel like suffering should be embraced and celebrated. (Note: I'm a practicing Catholic.) Sometimes things get out of hand. A seed we plant becomes a beanstalk that grows bigger and taller than we could have ever imagined. And while it's not necessarily the inventor's fault, it is necessary for us all to learn from instances of insanely rapid growth and always think about some of the possibilities for our projects that may be the antithesis of what we intended.

Technologists might *try* to anticipate the future the best they can. You can identify a few pitfalls by bringing together a group of experts from different backgrounds. I'm personally obsessed with doing this. I love to mix disciplines. Not only does it lead to interesting conversation, but it's also the best way to learn.

I've tried this in a roundtable to prove that it works. In 2015, when I was still at RAND, I wanted to prove that the power of an expert roundtable would lead to surprising insights. I had the honor of hosting Andy Weir, author of *The Martian*, a book about an astronaut stranded on Mars. To mark the occasion, I pulled together twenty-five experts from RAND working in non-space sectors. The room was a one-of-every-kind, human version of Noah's ark, with a clinical psychologist, an anthropologist, a political scientist, and twenty other -ists and -ologists. Our goal was predicting or identifying "critical technical, social, and political factors that are likely to be the key items as present-day humans endeavor to explore and inhabit Mars." As part of the ground rules, the group was allowed to assume that basic necessities like water, food, and shelter were already handled. I didn't want them to get bogged down into how to purify urine into potable water, but instead wanted them to think of all the problems humans might encounter or should consider when colonizing Mars. Weir had described the barren terrain of Mars in his book. We tasked our experts to combine their expertise and specialties to explore the challenges of colonization.

The experts wondered when the Mars colony would need currency and what the governing body of Mars would look like. They thought of everything from hydroponic farming to how Mars societies would deal with childhood obesity. They spent a great deal of time considering how mental health services would be provided. They discussed whether humans had the rights to colonize Mars, and what made us believe we could charge people for the privilege of visiting. They ruminated about class. Would the uber wealthy be the only people who could stake a claim on Mars? Would Mars become the next Dubai?

In the end, the team discovered so many hurdles and considerations that they concluded a colony on Mars should be a carefully constructed and considered entity decided upon with extreme sensitivity, many even voting against it. Andy smiled with amusement as he witnessed the non-mathematical challenges of space exploration unfold.

My hypothesis was proven true—that groups of experts from such a variety of fields can see and question a project that has previously stayed siloed within one set of disciplines. The wide variety of considerations and questions sparked during this discussion inspired me greatly. It led me to believe that a two-hour roundtable should be a requirement for every new business owner or new tech entrepreneur.

I have a feeling a roundtable might have identified some dangerous human behaviors or outcomes that would result from an app like Instagram invented to simply show off photos. Of course, I don't know. I only have hindsight. I understand that not everyone has twenty-five world-class experts sitting around waiting to talk about hypotheticals. Plus, humans are creative by nature. There is always someone who will think of a new use for a new thing. There are always the Banksys of the world who will hang bat-like from tall billboards and spray-paint their artwork. There are always people with Peter Pan syndrome who remember what it is like to turn a garbage can lid into a stage.

You bought this book to read it, but there's always a chance you will use it to make paper airplanes. Please post pics on Insta if you do, and tag me, of course.

This is all to say that you might be using technology in harmful ways, but that might not actually be your fault.

The internet can be a beacon of community, it can spark grassroots movements for change, it can gather people to send supplies to war-torn countries, it can teach an entire country how to do the same dance, and it can collect a GoFundMe retirement fund in a matter of days for an eighty-one-year-old whose Social Security benefits don't pay her bills. Yet, every time we go online, we sense that technology designers or corporations in cahoots with them are manipulating us in a myriad of ways. This may cause us to feel overwhelmed, powerless, or addicted every time we waste hours scrolling, knowing we should be doing something in the corporeal world instead. We may feel an undercurrent of remorse when we post something to our profile that isn't 100 percent true.

But that isn't really the technology's fault either. It is true, those factors do have the potential to dysregulate us every time we go online. Sometimes it's good to look at not only what's harming you but to also ask, "Why?" Are you using that technology in a way different from its original intended use?

Are you using Instagram to share photos or to compare yourself to others? Are you pouncing on opportunities to judge? Are you using Amazon to get something you need delivered because you can't find time to go to the store? Or are you using Amazon because you're bored and like to fill your cart late at night when you can't sleep? Are you checking your phone because you are expecting a message or because you're addicted to the dopamine fix you get when a new message comes through? I just checked, and I averaged just over three hundred phone pickups per day! Unfortunately, researching this topic did not make me immune.

I didn't write this book to shame you or berate you for checking your phone every twelve minutes or comparing yourself to strangers on the internet. That's just what we as a society do now. It might not have been the intention of the creators and technologists either, but here we are.

If you're feeling lonely, broke, or lost, it might help to simply consider the multiplicity of potential uses of your technology. Consider the two sides of each product—the potentially innocuous intentions of the product designers and the perhaps more duplicitous intentions of the companies who now own the rights to those products. Consider how you might tap into the potential for good in these products, to take control of your screen time. Are you spiraling into a vortex of Facebook clickbait ads directing you toward beauty products designed by companies deliberately playing into your insecurities? If so, what is a more responsible way to use Facebook to find the same types of products or even to examine your issues around these products? Can you message a few friends and ask them what products they use? Can you look for reputable product reviews in trusted magazines?

If you're a technologist, it might help to consider seating all those folks at your roundtables, and getting a discussion going about every possible outcome of your technology. Understand the potential for harm. Try to find a way to educate people about ways to use your technology responsibly, in ways that are aligned with your original vision for it. We don't necessarily have to surrender control to the highest bidder.

I'm talking to myself. I'm in the middle of building a groundbreaking technology that revolves around my favorite topic: stories. I'm building it with the best intentions. Hopefully, by the time you read this book, it will be in your hands already. I hope you'll be using it in all the ways I've imagined. I'm sure some of you will get creative and see other ways to use it. My grandfather taught me to embrace human ingenuity and multitudes of possibilities. I only ask that you use this technology wisely.

CHAPTER 2

We're in Trouble

I want to be a good mom, but I'm so scared, you know? It doesn't really feel like I have much control at all. Sure, I can give hugs and tell them I love them, but society seems to dictate so many things. I made sure to give them yellow and green so they wouldn't be bound by some blue and pink rules but guess what they ended up loving? Pink and blue. I do such a good job not body shaming myself, and guess what my eight-year-old daughter said the other day? That she's fat. And on and on. This world we live in has such a hold on who we become! Are my kids going to ditch their childhood imaginary friends one day for actual AI imaginary friends? Will their brains eventually be rotted away by microplastics? Will they become adults who can even function in this world? Oh man, I really hope so. I hope so sooo much. It's all I can do but hope so.

—Anonymous, audio-recorded story

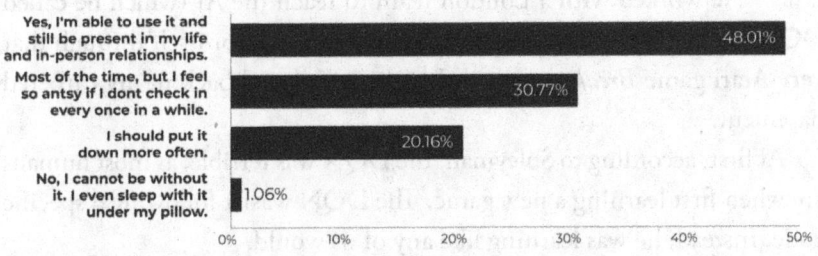

Do you think you have a healthy relationship with your phone?

- Yes, I'm able to use it and still be present in my life and in-person relationships. — 48.01%
- Most of the time, but I feel so antsy if I dont check in every once in a while. — 30.77%
- I should put it down more often. — 20.16%
- No, I cannot be without it. I even sleep with it under my pillow. — 1.06%

Represented sample of the US population—350 people performed March 2024

I feel lucky to be old enough to be able to remember a time before the internet and cell phones. I practiced cursive writing with my number 2 pencil until my black and white composition notebooks were full. I watched the flag wave while the National Anthem played on my family's television set before the station shut down for the night and the television screens turned to static. I henpecked the clunky keys on a Commodore 64 in my classroom. I printed homework using a dot-matrix printer with perforated paper. I made mixtapes on tape recorders. I found books using the Dewey decimal system, which we were forced to memorize. And yes, I was among the prehistoric human beings who witnessed the advent of the internet, often pacing around my living room while the screeching dial-up took ten minutes to catch.

In the '80s, I came home after school and played Atari, a prehistoric system for playing prehistoric video games on a big honker of a prehistoric TV set back in the day—as deep as the screen was wide. We often played *Breakout*, pushing the heavy weighted joysticks in our laps to control the game on the honker TV screen, bouncing a ball off a paddle, which knocked down rows of colored bricks.

Breakout is kind of like Ping-Pong in that you move the paddle to hit the ball, which then bounces from the paddle to the bricks, knocking the bricks down one by one . . . unless you're a skilled player and have figured out some better techniques.

In 2012, Mustafah Suleyman developed Deep Mind for Google, an AI system to train that would "exceed human performance at most cognitive tasks." He worked with a London team to teach the AI (which he called DQN, or Deep-Q Network) to learn how to think for itself through that very Atari game *Breakout*, which I had once played back in my carpeted basement.

At first, according to Suleyman, the DQN was terrible, as most humans are when first learning a new game. The DQN wasn't following a specific code. Instead, he was learning like any of us would.

One day, the DQN discovered a new strategy all on its *own*! According to Suleyman, "Instead of simply knocking out bricks steadily, row by row, the DQN began targeting a single column of bricks. The result was the creation of an efficient route up to the back of the block of bricks. DQN had tunneled all the way to the top, creating a path that then enabled the ball to simply bounce off the back wall, steadily destroying the entire set of bricks like a frenzied ball in a pinball machine."

This was a strategy most humans would never have figured out on their own. The DQN had learned independently. This was huge—an absolute breakthrough moment when the team realized that an AI agent could discover new strategies on its own.

I thought AI deserved a quick mention up top because it is now one of our biggest wildcards. Who could have predicted that our antiquated Atari games (one I played so much it occasionally cramped my hands) could have been used to teach a technology that is not human to think kind of like a human and then to go ahead and think better than a human? Who could have predicted that teaching Mr. DQN to learn could have been one factor that led to AI's rampant evolution?

Evolve it did. It totally did. In China there are newscasters (two, one fake man and one fake woman) who look like humans but are entirely run on AI that broadcast the news into people's living rooms. I just listened to a podcast about a guy who is cheating on his wife with an AI girlfriend. There are self-driving cars that relegate the humans sitting behind the wheels to "secondary drivers" who are not supposed to touch the wheel or the brakes unless the robot driver messes up (and by that time you are generally already doomed). And there is a myriad of everyday ways our pal AI mines for and often exploits our data for purposes over which we as humans have little or no control—which right now could include things critical to our humanity or survival—making real paintings that have meaning, landing a job, or getting approved for a loan.

AI became so normalized, so quickly, we don't even know how to refer to it through language. We can't decide if we should use a pronoun or a noun when discussing it. Is it a machine we should not try and anthropomorphize and always refer to as an *it*?

For purposes of this book I will define AI as anything that meets these three requirements:

1. It's autonomous (meaning it can do things all on its own including make decisions).
2. It can learn.
3. It's non-biologic (meaning it's not a human or animal or other kind of living being, and we created it).

To make it even simpler: AIs are machines we created that can operate without assistance and learn.

Perhaps more important (and more terrifying): We can't really wrap our minds around what the future of our spunky little circuitous genius disembodied tin will entail. We fear for humanity. Outspoken prognosticators like our friend Elon Musk are predicting that AI will take on all employment and "probably none of us will have a job."[1]

Other individuals (from conspiracy theorists to the sweet grandma next door) are proselytizing that AI will usurp all our artists, writers, and innovators—essentially removing all emotion and humanity from our paintings and books. Governments are fighting for mining rights to rare earth minerals all over the world that can be used for some portion of AI's guts or life support. Many are terrified that AI will take over. Their minds may even jump to the worst-case scenario—that the most potentially monstrous, more embodied AI may someday run amok, cause destruction, and even start wars.

I like to believe that is all pretty far-fetched, that the future isn't that shady. But the truth is we just don't know yet. We are in a kind of existential limbo. And I actually believe that we have the power to do something about it, but we have to act now.

Think about what this kind of thinking does to our kids.

If people of my generation who played games like *Breakout* (and sometimes even used hand squeezers to make our hands stronger to really work that joystick) are caught in a technological time warp, what does it feel like for kids now—those Gen Zers or Gen Alphas who never knew what life was like before three-year-olds were given cell phones?

My kids are caught somewhere in the middle of the new worlds. Their eyes were not always stuck to the flypaper of their screens. As mentioned earlier, they are lucky enough to have invented primitive, creative games. Remember, they played Peter Pan on trash can lids, and made office desks out of swings. They played all kinds of games where they used no props at all, and turned themselves into all manner of magical creatures, simply by using their bodies, their voices, and their imaginations.

I remember my daughter, Alaina, as a five-year-old playing "lions" with her friend. They'd rotate being the big fierce lion or the nurturing mother lion. They'd roll around on the living room carpet and roar and snarl at each other, then bash into each other gently on the floor. They'd purr and laugh and roll around. Once, they were playing and laughing while I was washing the dishes. Then suddenly there was silence—which is never good when you are dealing with five-year-olds.

I turned around. I saw both girls' expressions had changed. They looked hurt, scared, and it was as if some momentary emotional butterfly flutter had passed through them. Then my daughter looked at her friend—a precocious redhead who wore glasses.

"No, I don't want to play lions anymore," my daughter said. "Let's play cheetahs instead." To this day, I'm still not sure what was the difference between lions and cheetahs, but they sure did—and that's the point, isn't it?

When kids play games off-screen, they are low stakes. Kids are sensitive. They know instinctively when things are in danger of going the wrong way and care about each other's feelings, so they know how to shift them. Nowadays, the play disappears early on as kids shift to screen-centered communication as early as nine or ten, depending on their friend groups. Instead of Barbies, it's now Instagram or Snapchat. These platforms are not low stakes. You post one wrong picture, and you're canceled. The whole school hates you. One wrong comment and your bestie disowns you. Everything is big. Everything means something. How close or intimate can those relationships get if they can abandon you at any moment?

Jonathan Haidt, my favorite social psychologist and author of *The Anxious Generation: How the Great Rewiring of Childhood Is Causing an Epidemic of Mental Illness*, blames technology and social media (among other things like overparenting) for a host of issues within the Gen Z population. He cites the lack of play in childhood these days as one of the main factors affecting our kids' social skills, physical skills, and their abilities to properly judge risk.

Tweenhood is a time when you're wrestling with identity and trying to figure out what you stand for. I know I did. I switched up hairstyles and clothes every time I moved to another Air Force base, wondering who I should be this time and if it would earn me more friends. Social media platforms, however, require that you cultivate a persona and perform it for everyone you know. Just when you're trying to figure out who you are, you're forced to display it for everyone and wait for people to judge it—and judge it they do.

By the time they become teenagers, they are living much of their lives through their online doppelgängers, and many are not clear on who they are. The process of self-actualization that naturally occurred by playing "lions," creating extended families with their friends, confiding in them, and forging their identities is no longer an option. What happens when our social interactions are circumscribed to our screens? On the one hand the internet may be a way to find your tribe, to hand-pick that tribe from options all over the world. That might work for some, but it's not enough for most people. Social media platforms usually favor likes and followers instead of the humans behind those likes and followers.

Teens often post whatever they can to get more likes. It's posting to win instead of posting to connect. It's performative. It's empty. It's not real. And it hurts to not be real. Imagine just making small talk forever and ever with people who are lying. Doesn't that sound torturous? This is the opposite of what we as Homo sapiens literally need to thrive. We are not getting the nourishment our brains need from social interactions. We aren't learning how to be humans in relationships. We are losing our sense of self.

In this case, AI provides people online with friends who are literally fake and makes everything more confusing and less trustworthy.

Haidt calls out the addictive levels of comment validation. We don't just put our screens away and focus on the present moment anymore. Many teen girls, the demographic tortured most by Instagram, report putting it away for a while but that it never leaves their thoughts as they wonder who might be commenting while they're not looking.

It's a stressful experience to wait for a comment to appear before you can you relax in the knowledge that you're seen or accepted by your peers, at least for the day. Each day can be a roller coaster of emotions. Many teens report being on phones late into the night and then again early in the morning. There's sometimes no time away from the oscillation between ridicule and validation.

How can tweens try anything now when one hair out of place could mean they're ridiculed ? By the time they become teenagers, they are living much of their lives through their online doppelgängers, and many are not clear about who they are. Often, they don't get the support they need online but don't know how to seek it out or forge relationships completely offline. They aren't accustomed to getting by "with a little help from my friends." They don't often confide in their parents.

According to the CDC's American Time Use survey, American teenagers are spending almost 50 percent less time doing face-to-face socializing than they did twenty years ago.[2]

This is not only alienating, but also dangerous. For nine or more hours a day, kids have influencer after influencer telling them to be someone else, to cut themselves, to eat nothing, or to bully someone else. They have older men slipping into their DMs telling them to be sexy. They have a novel a day of data entering their sweet, malleable brains to let them know they should be someone else entirely. It hurts, it's confusing, and it's causing us great harm.

Each year, the CDC conducts the Youth Risk Behavior Survey, which asks ninth through twelfth graders questions about a range of health behaviors. In 2021, they released the trends over a ten-year period. NPR dissected the survey and revealed that 60 percent of teen girls had depressive symptoms in the past year, which is the highest ever reported over that ten-year period. Nearly one in three girls surveyed reported seriously considering suicide over the past year.

That is an *alarming* amount of kids who feel like they want to die.[3]

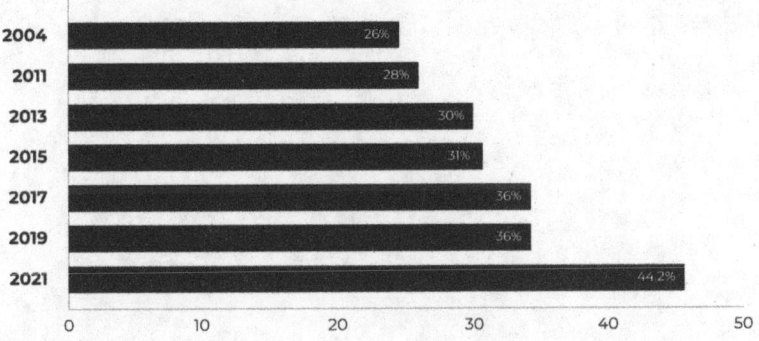

Credit: Derek Thompson, The Atlantic; data from the CDC.[4]

What does it say when every third teenage girl feels estranged enough from our society that she becomes preoccupied with self-harm or suicidal ideation? Feeling like you want to leave the earth is the epitome of loneliness—even if those who experience it generally note that it generally occurs while they are going through the motions of their daily lives, and nobody would suspect they were in turmoil. How noninclusive is a society that makes so many teenagers feel displaced or alien enough to take action to end their lives?

Is this damage irreparable? Are children born in an era where they are told technology that teaches itself may one day make any career or life purpose completely irrelevant completely swallowed up by the machine? Can it truly take away abilities previous generations had taken for granted: creativity, ingenuity, confidence, the ability to self-validate, confide in best friends, or think for themselves?

Shouldn't we claim more responsibility for the world we have left them?

Maybe this would be a no-brainer if we had more agency ourselves. Instead, their adult role models and guides (their parents and teachers) are also caught up in the circuitous, rapid misfire rat race. So many humans are trapped in our own cycles of technologically and economically induced pain. Will we be able to look up from our phones and do something about it?

CHAPTER 3

We're Lonely

I could feel my friend Jenny pulling away once she had a baby. And I get it. We were suddenly interested in very different things. I hadn't yet been shoved into the world of diapers, but I thought I would be someday. Then another friend had a baby and another. And I remained single. It's not like I wanted that. I was hoping to one day have a baby, but now I'm in my late thirties and haven't found anyone I'd want to have a baby with. Like, at all. And while I'm extremely sad about that, it's almost more painful to feel the loneliness of it all. It's like a silent goodbye from each friend, one by one. And sure they might text on my birthday, but it's not the same. And it's so hard to make new friends when you're older. I've taken pottery classes and whatever, but I'm missing that connection from people who really know me. It's so hard. I'm still here, but where is everyone else? I feel like I'm back in high school at the lunch table, and everyone found another place to sit, and nobody said goodbye.

—Anonymous, audio-recorded story

MEETING UNCLE CHRIS

There are a lot of ways to join the Armed Forces. I like to go all in, so I chose to enroll in the Reserve Officer Training Corps (the ROTC) while I was a student at the University of Virginia, studying chemical engineering. It was a way for me to be able to join the Air Force as a full-fledged officer right after college and immediately be a responsible and accountable adult. That is how I roll.

In the ROTC program, you go to a boot camp for officers after your second year of classes, known everywhere else except at UVA as your "sophomore year." Every cohort arrives clean, ready, smiling, pumped. While we do leave feeling empowered, we're also drained, dirty, and trauma-bonded. No matter how tough an officer may be, he or she still gets pushed to their absolute physical and emotional limits during boot camp, and the whole thing can make you want to cry. Crying together is an immediate way to forge friendships, and during my third year of college I found myself yearning for them and the connections I'd made there. I didn't realize it then, but I was lonely.

Some of the most painful and poignant moments of my young adult life occurred during the summer of 1998 boot camp in Shreveport, Louisiana. In many ways being a cadet in ROTC provided me with the kind of makeshift community I couldn't have gotten anywhere else. The physical and social rules were so stringent and conformist that I blended in a way I never could in places where I had more freedom. It was a relief to feel so constrained, yet supported.

One of the toughest physical tests at ROTC was a modified Norwegian foot march, which is a test of endurance for soldiers. We had to carry a backpack that weighed a whopping thirty-four pounds (almost a quarter of my weight—I was kind of scrawny back then), and march in combat boots for fifteen miles. I was in relatively decent shape but had not trained for the test. I have to admit I was so nervous—not scared of the physical pain but scared that I would humiliate myself by not finishing.

It was a tough slog, for sure, in Southern August heat, but by the time I made it to the finish line, clenching every charley-horsed muscle in my body and teetering on blistered feet—I was relieved to notice I was not the only guy with tears in his eyes.

Those tears forged an irrevocable bond, not with any particular individual but with anyone who has ever gone through the experience. I felt like that bond was missing in my engineering classes, so I accepted an invitation to visit a few of my boot camp buddies in Nashville, in the upcoming fall semester.

I couldn't stand the idea of being in the car alone for the long trip from Charlottesville, Virginia, to Nashville, Tennessee. One day, I walked into my chemical engineering classroom, announced my plans, and asked if anyone wanted to join me. The only one to raise his hand was this guy Chris who I didn't know that well at all. We'd hung out in a group a few times and we'd seen each other in classes. In fact, I wasn't even sure I liked him at all. At first, I felt uneasy about being in a car for such a long time with someone I hardly knew, but it would have been awkward to say, "Oh, I didn't mean you."

Just two days later, we jumped in the car and headed to Nashville. I actually remember the car ride more than I remember the trip itself. I don't remember who started to bypass the surface chitchat first, but we filled every one of those eight hours with some serious conversation. We spoke about our relationships with our fathers, our complicated feelings about organized religion, who we wanted to be, and who we thought we actually were. Obviously, we talked about girls and movies, too, but I mostly remember the deeper things.

Sharing so much with someone I barely knew made me realize how starved I was for connection—to get down past the surface level and really see a person and be seen. We definitely opened our hearts to each other, and by the time we got to Nashville, we each knew we had made a friend for life. Sharing that car ride with Chris really opened my eyes to how important such vulnerable connections are. It really impacted me. That kind of connection has always been hard to come by for me, so I hold onto it when I find it.

Chris is still like a brother to me almost thirty years later. He's since become a dual-MD (internist and psychiatrist) and is now my company's

Chief Medical Officer. To my kids, he's "Uncle Chris." Our friendship all started that day in chemical engineering class because Chris was brave enough to agree to hop in a car with a stranger on a long trip, and we decided to talk about something deeper than our classes.

THE CONSEQUENCES OF LONELINESS

I feel extraordinarily lucky I still have Chris as a friend, especially when so many people in America are lacking connection.

According to the nineteenth and twenty-first American surgeon general, Vivek Murthy, loneliness is a serious problem in this country, and half of US adults report feelings of loneliness. In 2023, Murthy released a seventy-five-page advisory, calling loneliness an epidemic, citing several studies proving its harmful effects as well as the benefits of social connection. This report claims that loneliness can be as dangerous as smoking fifteen cigarettes a day. And it can exacerbate all kinds of illnesses.[1]

Murthy's advisory reviewed a paper out of the University of Glasgow that analyzed data from over 450,000 participants, aged thirty-eight to seventy-three, in the UK Biobank database. The participants answered questions about loneliness or their feelings of being connected to friends or family. After about thirteen years, around 33,000 participants had died. The paper found that, regardless of the reason for the death, those who had no visitors at all had a 39 percent higher risk of dying than those who had daily visits from friends or family.[2]

I was shocked to see loneliness described as an epidemic, and I was even more shocked to read that patients without visitors in hospitals have such a greater risk of dying from whatever disease they're suffering. I wondered if I had anyone to visit me in the hospital. As far as I know, the answer is still a yes.

It may not be forever if I don't remind myself to care for my relationships.

I have Chris in my life, and I have a family I get to see all the time, but was I connecting with them? I work a lot. Chris and I talk a lot these days, but mostly about logistics and new business proposals. We don't take too many long car rides together. It's easy to put vulnerability to the side when you're busy, but how could I have let that happen when questions and stories are my literal full-time job?

I'm a researcher. My job is to be curious. I am supposed to ask good questions and be able to get people to open up, to talk to me, to get vulnerable. I am supposed to collect my friends' stories. Was I doing a good enough job?

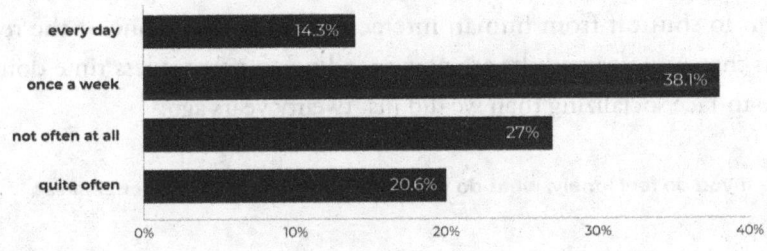

Maybe I was lonely. Maybe we are really all lonely. According to Murthy, at least half of us are. When we did our informal study for this book, it looks like many of us feel lonely at least once a week.

What's happening to our culture? Why are we so lonely?

When Harvard researchers asked adults who or what they think contributes to loneliness in America, technology (73 percent) topped the list, followed by families not spending enough time together (66 percent), people working too much or being too busy or exhausted (62 percent), and people struggling with mental health challenges that are hurting their relationships (60 percent).[3]

Yep, I can relate to all of these.

According to Mitch Prinstein, the chief science officer at the American Psychological Association (APA), "Within the last twenty years, the advent of portable technology and social media platforms [has been] changing what took 60,000 years to evolve."

The biggest of these changes deals with human interaction. Our brains need it. It's a human need to *not* be lonely. We learn from each and every human experience we have. Prinstein explained, "Numerous studies have revealed that children's interactions with peers have enduring effects on their occupational status, salary, relationship success, emotional development, mental health, and even on physical health and mortality over forty years later. These effects are stronger than the effects of children's IQ, socioeconomic status, and educational attainment."

When we shove a screen in front of our children's faces at a restaurant instead of teaching them to socialize with our family members, we might be changing more than their restaurant behavior. We might be teaching them to shut off from human interaction. It could be one of the reasons that American adults are now spending 30 percent less time doing face-to-face socializing than we did just twenty years ago.[4]

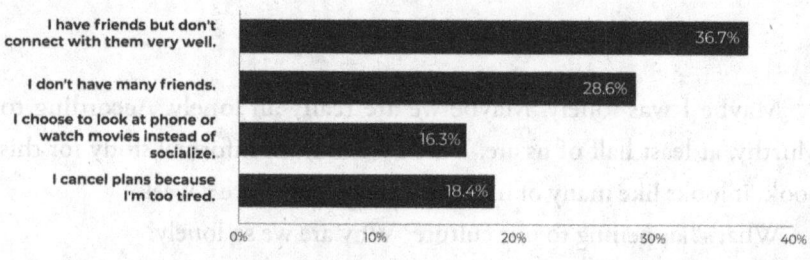

If you do feel lonely, what do you think contributes to that loneliness?

- I have friends but don't connect with them very well. — 36.7%
- I don't have many friends. — 28.6%
- I choose to look at phone or watch movies instead of socialize. — 16.3%
- I cancel plans because I'm too tired. — 18.4%

Loneliness isn't necessarily a result of being physically isolated from people. People can feel profoundly lonely even if they are living in a house full of family members, or going to work every day and BSing with their coworkers, or frequently visiting people who consider them friends. Superficial connections can often feel more emotionally taxing than forgoing physical connections at all. If you just talk about the laundry and nobody in your family really *knows* you, that can feel pretty isolating.

THE DISAPPEARING VILLAGE

If you're a Millennial or from Generation X and have Boomer parents, chances are you might feel abandoned by your village. There is a plethora of videos on social media about this topic. TikTok creator @BrittnieVH's video went viral because she captured the sentiment in a video that states that, for many of us, our family is "nowhere to be seen" and can't be bothered to be part of our village. She says, "We are our own village. Such a difference between us." Other people have shared their experiences with Boomer parents honestly online, creating a community of those who feel somewhat abandoned by parents who are more interested in their own lives than spending time with their grandchildren. Most share memories of how they spent so much time at their grandparents' houses growing up, as it was second nature to be with extended family. However, nowadays, in America at least, that type of a relationship is more of a rarity. Grandparents aren't as willing to babysit or be a big part of their grandkids' lives.[5]

Many Boomer grandparents will disagree, or they might have valid reasons for the change in village-like living (generational trauma or issues stemming from being raised by the emotionally lacking Silent Generation), but the fact is that many children of Boomer parents feel abandoned by those closest to them. Parenting has become lonely for many.

If we're no longer able to rely on our parents, that's a problem, especially because research also shows that we no longer depend on acquaintances much either.

As Marc Dunkelman wrote for *National Affairs* in his piece "The Transformation of American Communities," "Over the past few decades, technological, social, cultural, and economic changes have revolutionized the structure of American community."

He describes how society's current setup has allowed us to silo ourselves and essentially cut out acquaintances. He says, "The developments of the past few decades have served to weaken the ties that once bound local communities together. In their place, we are, on one hand, now choosing to invest more time and energy in keeping in touch with our closest friends and family members, and, on the other, in trading bits of information with people we do not know very well but who share some single common interest. As a result, the relationships that stand between our most intimate friendships and our more distant acquaintances—the middle-tier relationships that have long been at the root of American community life—have been left to wither."

He uses Saturn's rings as a metaphor to explain how we choose to commune with the rings closest and farthest, but no longer with those in the middle. Instead of catching a neighbor outside for a chat while grabbing the mail, we might choose to post that chat on Reddit instead, creating a community with those who'll never actually know us,

If that pool of close relatives and grandparents is dwindling and we no longer have a pool of semi-close acquaintances like neighbors, cashiers, and teachers, that leaves us with the Etsy seller we buy knitted beer koozies from occasionally. That feels pretty lonely to me.

"The result has been a dramatic change in the architecture of American community—with major implications for our economy, culture, and politics."[6]

It makes sense that Millennials may find comfort in community online in forums where people such as @BrittnieVH help to articulate a heart truth that people of one generation are feeling—identifying a problem that other people can relate to and post about. The posting is a kind of public confession, a cry for understanding and support. The fact that we are turning exclusively to anonymous people online who we don't know for support on this issue totally tracks. But it doesn't necessarily solve this issue in "real life." We're all out here posting. But we're still lonely.

HOPE IRL

The internet does occasionally help us organize collectively to take care of each other. There are certainly many makeshift communities that have been forged online to help others—often as a response to societal problems or crises—for example people who wound up sewing masks or delivering food for their neighbors during the COVID-19 global pandemic. Or there were those who organized GoFundMes and supply drops for those who lost their houses in the 2025 California fires. People organize to secretly leave new blankets and socks near encampments where people who live on the streets gather.

There are groups devoted to random acts of kindness who give out flowers to strangers in train stations or those who salvage bread from bakeries at the end of the day and redistribute it to people who are hungry.

Conversely, we are more collectively distracted and isolated than humans may have been in the past. People used to talk to each other in bars, coffee shops, and restaurant counters. Now we sit in those same places and stare into our phones.

What would happen, I wonder, if we denied people access to those phones?

The Offline Club hosts digital detox retreats and runs a pop-up café in Amsterdam where customers have to check their technology at the door. Footage of the coffee shop shows deeply satisfied people doing old-fashioned things—crocheting sweaters, reading, typing in journals, talking, laughing. If the video is any indication, nobody seems twitchy or distracted—the way people tend to get when technology has entered the room. Instead, there seems to be a deep peace, a free rhythm and deeper purpose to their interactions—than there seems to be when people are turning from their conversations every two seconds to check their text messages or looking down at their screens to catch another can't-miss text. Who knows, this video may have been taken during a peaceful five-minute interlude in the café. The patrons may really turn into nervous wrecks after twenty minutes in this technology-free zone, looking around the room like they lost their best friends or are grieving the loss of a cherished pet.

On the other hand, what if it's true? What if there were ways we could de-technologify our interactions every once in a while, return to the days when people were content to exist in the world physically in real time, when conversations ran deeper without all the interference? Is there a way we could replicate this same experience for our children?

British Journalist Decca Aitkenhead sought to find out.

S'MORES AND STORIES

A group of ten teenagers, ages thirteen to fifteen, sat around a campfire at night in the woods, laughing and telling stories. The sound of pine sap sizzling in the dense wood punctuated their laughter. Their bodies were silhouetted by a moonlit sky.

If this were a scene from some 1980s teenage film, nothing would seem strange about it. There was one thing missing from this 2025 scene. Can you guess what it was?

That's right, none of the kids had their cell phones. Or rather they had cell phones, but they were crude, prehistoric, black and white cell phones that didn't connect to social media. They weren't smartphones; they were dumb phones.

British Journalist Decca Aitkenhead had gotten the kids to ditch the phones as part of a sociological experiment, in which she challenged them to unplug from their smartphones for a month. The kids (who included her two sons—Jake, fourteen, and Jody, thirteen at the time) had agreed to join the study, after they were bribed with financial incentives—part of Aitkenhead's concession to recruitment challenges.

Some participants admitted they were excited to try the experiment because it would give them an excuse to take a vacation from the pressure of social media. One of the girls, a fifteen-year-old named Edie, had tried to do a TikTok detox once before, "Because it steals all my time, and that makes me feel [like] shit about myself."

"If you follow every trend, you get called a basic bitch. And if you don't, you're a weirdo. There's no escape, because your social popularity is totally linked to your social media. So if you don't post, you get made fun of. But then if you lip-synch to the wrong song on TikTok, you get made fun of for the rest of the year," said Rose, a fourteen-year-old with an edgy style who looks cooler than I've ever been.

While the girls had a tougher time with the experiment, the boys said they read more, connected more, and slept more. In an interview weeks after the experiment, Aitkenhead's son reported to having kept his screen time down by at least 25 percent.

The unsupervised camping trip was the pièce de résistance of the experiment. They had to use real, printed maps to find the location, and they ended up getting lost but having a great time. They talked, sang, played guitar, made s'mores, told stories, and nobody scrolled.

Aitkenhead arrived at the campsite to pick them up and found a scene out of a John Hughes movie. "At noon on Sunday, campfire embers are still glowing in the heavy drizzle, so there clearly hasn't been a lot of sleep. Excitable chatter babbles out of the boys' big family tent. I peer inside to find [one of the girls] sitting in a large cardboard box, the other five snuggled in sleeping bags, and all wearing expressions that banish my doubts. The mood is electric."[7]

She recorded other responses on that magical day:

Talking over each other: "Oh my God, so much fun!" "Sooo much fun." "The fire was amazing. We got a ton of logs from the wood, and we made s'mores, and Isaac and Rose failed to play the guitar." "We didn't fail!" Rose is indignant. "I played Radiohead!" What was the best bit of the trip? "All of it."

Not everyone has the opportunity to join an experiment where their peers and friends put down their phones to enjoy a real connection. I know just one camping trip won't cure loneliness. It helps, though. When we stop scrolling for a bit and spend enough time with each other to share stories from our pasts and create new stories together, to figure out how to light a fire or get lost on the way to an adventure together, we may find we have a lot of fun. Those things wouldn't be easy to do if we were stuck in our phones or focusing on documenting it all "for the 'gram."

As an adult, it's easy for me to look at the next generation, judge them, and think I've got it all figured out. Just put your phones away and you won't be lonely, you silly teens. However, the solution is not all that straightforward. Loneliness has seeped into all our consciousness. I couldn't even describe my own feelings when I began pondering my own loneliness. I don't have it all figured out. I don't think many of us do. What I do know is that most of us could use more connection and less loneliness.

Once I realized this, I called many old friends and spent a few months catching up with them, asking them questions, and hearing their stories. Who knows, I might even host a story night for the neighborhood this weekend.

CHAPTER 4

We're Broke

I just moved my family to Portugal. I couldn't afford to be in the United States anymore. The thing is, I made great money. My wife and I did well for ourselves. On paper, it seemed like we'd be fine. But prices aren't fine. Childcare for our three kids was more than our mortgage. And then I would get so scared every time we went to get groceries. One Costco run and I was down $600. My wife and I found we weren't spending any time together because just going out on a Friday night would mean we pay the bill for the meal and the theater or whatever we ended up doing and then at least $100 for the babysitter. Plus, her cab back home in the dark, another $60. It was painful to do our bills every month. And don't even get me started on our health insurance. We are both self-employed, so that was a huge expense. On paper I was rich. In real life, I could barely stay afloat, and I was dying of anxiety every time I had to go get gas. Now, in Portugal, life is so much more relaxing in many ways, but especially financially. It's amazing to go to the store for groceries and not be worried about whether what I have in my account will cover it. That's a feeling I haven't felt in so long.

—Anonymous, audio-recorded story

In the years following the COVID-19 pandemic when inflation was at its peak, I could feel the gap between my friends who were uber successful (I may know a handful of these folks) and those who were struggling (I know a lot more of these folks) intensify and alter how we gathered together. People who used to be able to afford a few dinners out per month or whatever groceries they wanted to buy in the store suddenly felt strapped. They felt their comfortable financial status slipping. Many people were laid off. I saw more friends shy away from invites to big dinners where they knew the huge bill would be split among us or decline more invites that required significant financial commitments. In the meantime, other friends said "see ya laterz" as they jumped on private jets to go catch the Super Bowl. It didn't seem fair.

I've had the good fortune of being good with money. Not "good" in a way that means I'm a gazillionaire. I don't come from money, nor do I drive a McLaren. That's what I mean by "good." I would rather never drive a McLaren ever, or even a hybrid Porsche, even if I think they are sweet. Instead, I weigh my options and think about how much I could invest if I don't buy a McLaren or a hybrid Porsche.

I would label my financial choices as pragmatism rather than frugality, although my wife might disagree. She has, on many occasions, ordered me to buy new clothes when I'd worn my perfectly good ones decades after they had gone out of style. If it were my choice, I'd wait out cargo shorts until they return to runways. This thought process has served me well in times of hardship. I have savings. I'm used to making tough decisions when it comes to spending. For example, when it's time to buy a car, I really do my research and ask myself why I want that specific model and if I really need it.

When I was a first lieutenant in the Air Force, in 2003, I bought the first and only brand-new car of my life. I still love that Toyota 4Runner, which sits in my driveway several decades and 204,000 miles later. I have driven it through endless hours of LA traffic. It's made it across the country on five occasions. It's been to over two dozen national parks with me. I still use it for most rides to the grocery store.

That truck has brought me so much joy. At one point, my 4Runner was worth more than I owed on it, so I refinanced it to pay for my wedding, kinda like a truck equity loan. While I was paying that loan off, I was also paying for graduate school and paying to support my young children aged two and six months at the time—all on my modest military salary. And I pulled it off!

I can go without. I can downsize. I can get creative about where every dollar will go. When there's a recession or inflation or some big financial catastrophe, it hits me a little less hard than some of my friends and colleagues because I'm already used to the recession I've created in my own head.

I'm not sure my practical financial mindset is always the healthiest practice. I don't take vacations unless I'm visiting family. I don't treat myself to frivolous things. It's possible this mindset depletes my life of joy. Yep, this has been intimated more than once. It sure helps me in America when inflation is high. This was demonstrated during the COVID-19 pandemic when costs skyrocketed for everything from food to utilities (which have never quite recovered since).

Those prices intensified the gap between the haves and have-nots. I'm not an economist, thank goodness, but I do know enough to realize that prices don't recover if demand stays high. Companies aren't that altruistic. It's enough to make a lot of us clench our wallets tightly and vote in an election based on the promise of lower prices.

PUBLIC PERCEPTION

How do you feel these days? Rich or poor? Anecdotal stories from my community and evidence from the Federal Reserve demonstrates the middle class is shrinking and has been shrinking for a while. You're either rich or poor. Or (at least) you *feel* rich or poor.

While just 10 percent of the US adult population is raking in the big money, most of us feel pretty poor. In 2024, the St. Louis Federal Reserve Bank's Institute for Economic Equity published some numbers that took a while for me to process. The agency provides quarterly data on racial, generational, and educational wealth inequality based on the average US household wealth.

They define wealth as "what a family owns less (or minus) what they owe." That includes cars, property, homes, savings, investments, and anything owned. They subtract any debts owed, like credit cards or mortgages. Here are the most surprising facts:[1]

- The top 10 percent of households had $6.9 million in wealth on average. As a group, they held 67 percent of total household wealth.
- The bottom 50 percent of households had $51,000 in wealth on average. As a group, they held only 2.5 percent of total household wealth. Of this group, some 9.9 million families (about 7.5 percent of families overall) had negative net worth, meaning they were in debt.
- Half of all Americans have an average of $51,000 in wealth and 40 percent of the population falls between having $51,000 of wealth and $6.9 million in wealth.

I'd say that's a pretty large gap. It's more like a ravine that's widening by the minute. According to the Urban Institute, Lyndon B Johnson's think tank that researches inequities, "Wealth inequality is higher in the United States than in almost any other developed country and has risen for much of the past 60 years."

If 10 percent of the people own 67 percent of the wealth, that doesn't leave so much room for middle-of-the-road families to get a share.

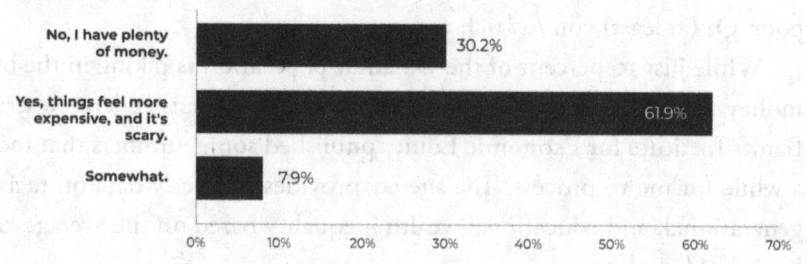

This is a lot of numbers, but it's important for me to see the disparity like this. Most of us feel financial stress every day, but seeing it in numbers validates how common it is. We don't just need another side hustle. Most of us are struggling.

It doesn't feel good. Living paycheck to paycheck is quite uncomfortable and can feel like survival mode. It's not easy to be joyful in survival mode.

Bernie Sanders's concept of the wealth gap being an American disgrace has endured. His Robin Hood wealth redistribution is still a fantasy.

Donald Trump, who became president in 2016, may have exacerbated the gap giving more tax breaks to the wealthy. He cut the corporate tax rate by 14 percent, which is a huge amount of money that corporations didn't have to pay. Billionaires and corporation owners got to save even more money. Great for them! Unfortunately, this created a bigger divide between rich and poor.

Conceptually, tax cuts for the rich are justified by the theory that those corporate savings would "trickle down" to poorer people, providing them with higher wages and better prices. It assumes that big corporations—like Amazon, for example—would say, "Hey! We made so much money this year. Let's give everyone raises!" It's difficult to find solid examples of where trickle-down economics actually resulted in practical, financial gains for poor people.

Part of our issue is that wages have been fairly stagnant. In 2018, *TIME* magazine ran a cover story. "I have a master's degree, sixteen years of experience, work two extra jobs, and donate my plasma to pay the bills. I'm a teacher in America." This was the headline for Katie Reilly's investigative feature,[2] which included interviews with several teachers who had to take on more than one job to make a wage that paid for just their basic needs. She wrote about how teacher wages had stagnated and most educators in 2018 were still making what teachers made in 1990.

That doesn't seem fair when the average price of bread in 1990 was $0.75, and I just paid $7.50 for a loaf. It was a loaf of Dave's, but still. Actually, in 2024, the average price was $2.54 for a pound of white bread—still quite a bit more than $0.75 when wages haven't risen. Teachers were making an average of $31,367 in 1990 (which would be about $65,000 adjusted for inflation). In 2025, when teachers have to do a lot more in their classrooms including conducting open shooter drills, they make an average of $50,000. That's even less than they made in the '90s.[3]

So, while the price of bread has risen, avocados are almost $4 each, and eggs are $5 per dozen, teachers have barely seen a rise in pay. Prices are going up, the rich are getting richer, and wages aren't budging. A 2021 Bernie Sanders blog post illustrates what it feels like to live in a society that's divided by such a large ravine:[4]

> Today, half of our people are living paycheck to paycheck, 500,000 of the very poorest among us are homeless, millions are worried about evictions, 92 million are uninsured or underinsured, and families all across the country are worried about how they are going to feed their kids. Today, an entire generation of young people carry an outrageous level of student debt and face the reality that their standard of living will be lower than their parents'. And, most obscenely, low-income Americans now have a life expectancy that is about 15 years lower than the wealthy. Poverty in America has become a death sentence.
>
> Meanwhile, the people on top have never had it so good. The top 1% now own more wealth than the bottom 92%, and the 50 wealthiest Americans own more wealth than the bottom half of American society—165 million people. While millions of Americans have lost their jobs and incomes during the pandemic, over the past year 650 billionaires have seen their wealth increase by $1.3tn.

Poverty makes ordinary daily life extraordinarily stressful. Living paycheck to paycheck, wondering if you'll be able to feed your family healthy food, or enough food, or pay your rent, and finding it absolutely impossible to get ahead, can be maddening.

So what are we doing about it?

The Guardian interviewed American grocery shoppers, asking them how they are dealing with rising grocery prices, and they had some good strategies. Many had to change their diets completely, cutting out expensive beef or serving their families whatever is on sale. Some report spending hours online each week looking for coupons and sales. Others report giving up everything good (like sweets and treats), and others say they go to three separate stores per week to get the best deals.[5]

If the mental overload of preparing dinner weren't hard enough, now we have to clip coupons, drive to three stores, and search for good bean recipes just to eat. That's a lot of work to stay afloat. Not having enough can wear on you, especially when there's a growing sentiment that it won't get better. Most Gen Z kids (my daughter included) don't really trust institutions. A Gallup poll from 2023 reveals they feel almost no faith in Congress, the news, the president, or large tech companies.[6] After feeling scorned by the corporate giants who've hurt the environment, littered plastic everywhere, and don't pay fair wages, they don't really have faith that corporate bosses that received all those tax breaks will somehow change their minds, take less profit, and decide to pay their employees a livable wage. They don't see a light at the end of the tunnel, and that can weigh heavily on our already lonely minds.

We're in despair about the future. The cost of food staples, rent, and utilities are rising but wages are stagnant, people are living paycheck to paycheck, the homeless population is rising, and many of us are feeling pretty poor (because we are poor when compared to the billionaires that hold two-thirds of the country's wealth).

So, what do we do? We zone out, we doomscroll, and we consume what we can—all the content. All the news.

And we *buy more*! We buy more, even when we don't have the money. Forty-two percent of Americans admit to living beyond their means. The *US News & World Report* explains the average American household in 2024 had around $6,100 in credit card debt.[7]

The average credit card interest rate is hovering at 28 percent. Americans continue to compound their debt to pay for both their needs and their wants. In a recent hearing on competition in the credit card marketplace, Sen. Josh Hawley pressed Visa and Mastercard representatives for taking advantage of Americans (who owe a combined $17 trillion to major credit card companies). That is a whole lot of debt.[8]

Sometimes we buy to keep up with the neighbors. Sometimes we buy things just to quell our sadness or help us feel like we can control a small shred of something. The University of Michigan's Ross School of Business conducted a study titled "The Benefits of Retail Therapy."[9] They found that shopping (and even just filling an online shopping cart) enhances feelings of personal control. And who needs to feel more control than those who also feel in despair about their finances?

Of course, we don't all just fill our metaphorical carts. We swipe that card! We look at those fake pristine houses on Instagram and convince ourselves that we need to buy a complete room makeover.

My financial pragmatism (or what my wife, Eileen, would call "cheapskateness") has saved me from having too many bad experiences buying big-ticket items online or straight off of Instagram. However, I am a sucker for backpacks that look great in photos but have pockets in all the wrong places and are difficult to return because they have "lifetime warranties," and "I don't like the pockets" is a hard sell for returns. I have far too many of these flawed backpacks in my closet because I may need that configuration of pockets one day and can't bear the thought of throwing them out. Somehow the stuff is much better off sitting in my closet than in a landfill. I've accepted my weird habit, although the rest of my household jokes relentlessly about it.

"Hey, you; you really need this thing," those influencers say. We buy that thing. It makes us feel really good for a while. It *works*! We feel in control. We can make purchases. We can choose the shipping options. We aren't actually at the mercy of any paycheck. Cue our evil laughter. Then—haha—we win! We get the package. We have the new thing! *We are in control!*

But when we then run out of money for household supplies or we can't pay our credit card bills, we can have regrets about that new purchase. We ruminate about how stupid that purchase was. We might not even like what we bought after a few weeks. We kick ourselves. Yet, our phones and computers are designed to make us prone to buying stuff we don't need and can't afford again and again and again. And we do! According to Adtaxi, 68 percent of adults made a purchase directly through a social media platform in the past month.[10] Amazon receives about twelve million orders per day.[11]

Overconsumption is such a problem in America, there are now minimalist influencers urging you to not buy things. It's a trend. Creator

@katiaachesnok on TikTok, among many others, creates content in an attempt to *de*-influence viewers. She reminds you of all the things you do not need, from Amazon deliveries to lunchtime drop-offs. "I don't know who needs to hear this, but you do not need anything from DoorDash today. You do not need anything from GrubHub today. You do not need anything from UberEats today. You have food at home," she posts. This is the kind of content and storytelling I happen to love.

Overconsumption is not completely our fault. Psychologists and technologists are working together to keep you coming back to spend more, more, and more.

"It's called persuasive design, and it's 'all about money,'" says Richard Freed, author of *Wired Child: Reclaiming Childhood in a Digital Age*. Freed sat down with *Vox* to explain the practice of persuasive design.[12]

Developed by B. J. Fogg, a behavioral scientist at Stanford, persuasive design is an entire field of study based on research that proved that with some simple techniques, technology can manipulate human behavior. Freed says that Fogg's persuasive design formula requires three things to keep users coming back: motivation, ability, and triggers. In fact, tech companies hire psychologists to ensure that formula is being met. There are people who specialize in your brain's behavior spending their entire careers thinking of ways to guide you back to their products, and buy them, without requiring too much conscious effort on your part.

Why? Freed says, "Time spent on social media apps means more people will be looking at ads longer, and that will increase their revenue. It's an attention economy, and it's the job of these psychologists to make sure people look at these things for as long as possible."

We live in a country where half of us make up just 2.7 percent of the wealth, yet we've got a whole litany of psychologists persuading us to come back to the sales for just a little bit longer and spend a little bit more money. One corporate psychological tactic that keeps us spending in an unending loop is the automated subscription. How many automatic subscriptions have you got going right now? Most newspapers, workout influencers, and investment platforms use subscription-based models. It serves these companies well, trapping you into a contract that might not be in your best interest. Every month they get to charge your credit card. They're hoping you forget all about it. Many of us do.

Subscription models are so rampant that there are now companies you can pay (potentially through a subscription but I haven't checked) that are dedicated to helping you find all the subscriptions you signed up for and forgot about to help save you money. Adobe and Microsoft have changed their models so that now you can't even buy Photoshop or Word anymore. You must pay them monthly in perpetuity for something you will never own. This benefits shareholders but it isn't much value to you, the customer who just wants to save a few bucks.

It is nearly impossible to try and outsmart the system to save money when predatory brands are always thinking of more ways to take more from you and line the pockets of their CEOs. I think it's important to remember: Meta (holders of the keys to Facebook, Instagram, and WhatsApp) reported over $40 billion in earnings in each *quarter* of 2024. That's $160 billion in just one year made from your content, your eyeballs, your clicks, and the ad revenue of companies pushing ads to get you to sign up for subscriptions. Their big bucks are all thanks to you.

Sanders reminded us of this on his Fight the Oligarchy tour in early 2025 with this staggering fact: Three people have more money than 170 million Americans.

The three people he was referring to are Elon Musk, Mark Zuckerberg, and Jeff Bezos. Those three men are making so much money from our data, our eyeballs that scroll, and our purchasing power, while so many of us are in debt or struggling while waiting for our next paychecks.

CHAPTER 5

We're Lost

I was running to the train because I'm always in a hurry, but I honestly wasn't in a hurry. Work was over and I had no plans. And then I heard this violin playing. And I stopped in my tracks. Like, it was so weird. It's so not like me to pay attention to street musicians.... I don't know why... I just never do, really. I live in Manhattan and they're everywhere and somehow I just learned to ignore them.

But this one violin stopped me in my tracks... like I said.

It was this woman. She was around my age. And she was *good*. I'm standing there, jaw open, in total shock about how good she is. And how much she *loves* what she's doing. She did not care who was watching, and I don't think she even cared if she made any money. She just wanted to play. She closed her eyes and stood there playing her heart out. Sometimes her body shook because she was so into it. And I was mesmerized. I was just so moved that she obviously *loved* what she was doing. You could just tell. And it sounded so good.

Anyway, I sat down, and I watched her play at least four more songs. And this is crazy, but I cried! I was so moved by this stranger that I cried. And they were happy tears. I was so happy for her to know what she loved. And a little jealous, I guess. She found her thing, ya know? What's my thing? When will I ever find my thing?

—Anonymous, audio-recorded story

Some conceptual version of the phrase "Know thyself" has appeared around the world since time immemorial. The Greek traveler Pausanias noted the phrase inscribed on the Temple of Apollo at Delphi. It was also written on the Luxor in Ancient Egypt a thousand years before that. It appears in Sun Tzu's *Art of War* from Ancient China, the Hindu Vedas, and even Confucius emphasized personal governance, which implies self-knowledge. Getting to know ourselves is an idea that spans cultures, eras, and continents. Yet it is not that easily translatable or articulable into a concept that everyone understands. We still haven't mastered it.

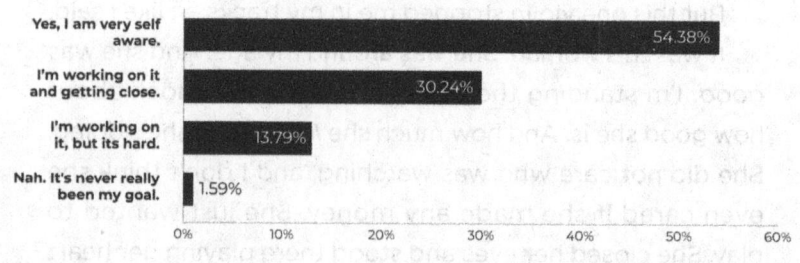

How, after all this time, do we still have so much trouble getting to know ourselves? Why did these ancient thinkers think it was so important?

Humans are the only species obsessed with creating a public-facing self. (At least, we think so. Elephants could potentially really care about what others think of their trunks, but it doesn't seem like it.) Perhaps that is the reason we get so lost. During the first part of our lives, most of us learn to create our public selves. Then we go through another period in which we finally realize that public self isn't serving us, and we try to uncover what's below.

To me, "knowing myself" is about getting to know that public-facing part and then tearing it down to reveal what is real, true, or honest about my wants, needs, and actions. We have an innate need to understand who we are behind all the superfluous things we use as distractions: our clothes, positions in life, job titles, place within the sibling order, where we were born, sports teams, cars, amount of money in the bank, etc.

I once believed I had a pretty accurate sense of the real me—beneath all the labels and doppelgängers. I thought that my sense of self wasn't impacted by other people's perceptions of me. Then I agreed to be the guinea pig in my team's sociological experiment.

WEST COVINA, CALIFORNIA, 2024

I sat in a chair in the cold room, the fluorescent lights shining in my eyes. I looked at the blue, curly wires hooked to the electrodes suctioned on my forearms. I tried to remain calm, to keep my vital signs as steady as possible. Part of me was amused, being hooked up to a polygraph machine like a suspect in a hard-boiled crime drama. I never imagined I'd ever be connected to lie detector wiring. I volunteered to be a human guinea pig for an experiment to determine if I would recognize myself when I was distilled down into bits and bites, patterns and trends, sentiment and behaviors.

The test was conducted by my own team to research how well humans *really* know themselves as we pave the way for our new AI-enabled technology. The technology we're creating is meant to helps humans know themselves better, and we needed to prove that our work was efficacious. Even though I knew my answers weren't being reported to police, I was nervous. Cold beads of sweat gathered around my temples.

"Are you in a good mood after you work out in the morning?" Pete, the polygraph operator, asked, his large hands resting on his dad belly.

"Yes," I responded, thinking, *Always*.

Pete made a confused face, and I didn't know if I had responded wrong. "Are you anxious after work meetings?"

"No."

"Do you feel like your sense of style represents who you are?"

"Yes."

I wasn't lying. I honestly believed all my answers. But I wondered if that's how serial killers also think when they're taking these tests. Finally, after several questions and with my T-shirt now drenched in sweat, I was unhooked from the scary wires. My colleagues and I went to the back room to analyze the results. I waited, patiently and slightly embarrassed. Lie detector tests were no joke.

I knew what my colleagues were doing. They were comparing the answers I had given with the data from my recordings. Every day over the course of the previous eight weeks I had been recording spoken journal entries about my mood, experiences, or encounters. Some recordings were a minute long and some lasted as long as ten minutes. I recorded my reactions to everything: a disagreement with an employee, a mountain biking trip, and an anxious presentation I made to a room full of investors. I emptied all my feelings into the microphone. This felt comforting to me since I've tracked millions of data points on myself using head-to-toe MRIs, every wearable in existence, and by studying biomarker details as they relate to my genetics. Talking into my phone's mic was easy. Every leadership book will tell you to keep some boundaries between you and your employees, but there I was entrusting my deepest thoughts to them. This was no privacy whatsoever. I might as well just have invited them into the bathroom with me.

My colleagues Katie and Chris had distilled all my recorded information into data points and composed statements, which I could objectively confirm or deny about myself. Later they came back to present me with the results of my polygraph. I passed with flying colors. I felt pleased with myself. I could finally stop sweating. However, I noticed Chris was grinning in that peculiar nervous way he had.

"Bill, the thing is, you absolutely, unequivocally believed every answer you gave on the polygraph. But in many cases, the answers did not match the recordings."

I smiled back. This was what we wanted.

"You hate working out in the morning!" Katie said. "You go to that boxing class you claim to love and then right after you record your most anxious rants."

I thought about that and realized they were totally right. They showed me the comparative results of the data they pulled from my recordings versus the results of the polygraph test. Basically, they proved that I don't know myself as well as I thought.

These were actually the results we were hoping for. We wanted to prove that we could help the humans that didn't know themselves well. I always figured that it was *other* people who didn't know themselves well. I go to therapy. I read a lot. I consider myself intelligent. How could I not understand the most basic things about myself?

I was happy to get the results we wanted: that most of us walk around quite oblivious to who we really are. However, I admit I felt disappointed to be just like everyone else. I seriously thought my rugged Patagonia jacket exemplified my sporty style, but in reality I am not as sporty as I think I am.

I'd suspected I wasn't 100 percent attuned to my experiences. Being so utterly misaligned with my emotional core was embarrassing. It turns out, my lack of self-awareness is not that unique. We all tend to say "fine" every time someone asks how we are. Maybe it's because we think the person asking doesn't really care—or maybe it's because we don't actually know the answer to that most common question. It's actually difficult to really live authentically within the confines of technologically advanced, fear-driven, socially convoluted cultures like ours.

I was also amazed by my results simply because I am a prehistoric adult who spent the first several decades of my life without social media or much internet, that I existed during a time period when our brains were more fully engaged with the corporeal world. I thought I had some innate sense of myself that predated this fragmented era. I individuated in the decades when we had actual dinner conversations and shot the shit with strangers at bars.

Well, if it turns out that I am deluded by my better, fitter, morning boxing ring, outdoorsy fashionista online self and/or some convoluted version of myself I pieced together in my imagination after having been ad zapped too many times—what happened to the kid who never knew anything else?

THE KIDS ARE ALL RIGHT. MAYBE?

There is usually a period of time during adolescence when we have a chance to experiment with our identities. We try different hairstyles or join different friend groups at school. We see what works while we create both our real and public-facing personas. At least, that's how it used to be. Before social media.

Since so many kids are now scrolling during this experimental phase, their experiments might not be coming from a genuine place of curiosity. Instead, they're being led by influencers telling them who they *should* be. It can be detrimental to their process of self-exploration. The content they are consuming can be harmful.

If you partake in your share of scrolling, you might be aware of a TikTok trend that surfaced around the beginning of 2024. TikTokers and Sephora employees were catching a new crop of customers in Sephora locations throughout America: ten-year-olds.

The "Get Ready with Me" (GRWM) video trend began in 2011 on YouTube and hasn't let up. Each one features a person looking into the camera as if it's a mirror and talking about a variety of subjects while performing their skincare and makeup routines.[1] I can't explain the enormous popularity with data, but I can confirm they are fun to watch. It's like seeing a before and after photo come to life. We watch someone transform right in front of us, almost while they're looking into our eyes. *Vogue* has hosted a captivating series of GRWM videos featuring celebrities getting ready, and it feels like a treat to take a peek into their bathrooms and watch them apply sometimes twenty products to their faces and necks. There have been so many of these videos (as of 2023, the #GRWM hashtag had more than 165 billion views on TikTok) that they must have snuck into the tween content algorithm.

In early 2024, girls as young as eight and up to fourteen began copying these adult routines and posting videos of their elaborate morning routines that included serums, under-eye masks, and retinol treatments. Then . . . as trends tend to do . . . it got out of hand. More and more very young creators began to post their ten-step skincare routines full of products their youthful skin did not need, and which could have damaging long-term effects.

Subsequently, IRL, Sephora locations nationwide were bombarded with young kids who couldn't afford to buy those ten products coming in to steal or mess up the tester products in stores. A video that went viral showed one tween having an absolute meltdown in front of a Sephora location because her parents refused to buy her an expensive moisturizer. I think it's safe to say that not one of those ten-year-olds would be aware of those skincare products without social media. I have a daughter in college now, and I'm so grateful she missed this by just a hair—a non–product laden hair.

Some people were outraged. Some people were shocked. Everyone had an opinion. Some blamed parenting. Some said it's just the newest trend for ten-year-olds who no longer wait in line for Cabbage Patch kids and pine for Drunk Elephant Vitamin C Serum instead.

I could relate. In the '80s when I was five or six, I decided I wanted a Cabbage Patch Kid, a strange, exorbitantly priced (for those days) doll with the name of their designer—Xavier Roberts—hand-stitched on its bum, which came with its very own birth certificate. These were the pre–Black Friday 4:00 a.m. stand-in-line years when the stakes on the few highly hyped trendy high demand items like the Cabbage Patch were an anomaly. My sister had one, the neighbors had them, and I wanted one. My parents could only find a baby girl version of the doll and had to fabricate a birth certificate, dress the doll in my baby clothes, and present it to me as a boy. I was so happy and never knew the difference. (Thanks, Mom and Dad!)

If the rush to collect beauty products is just this generation's Cabbage Patch craze, I guess I get it. But if kids are thinking they need these goops and creams in order to be beautiful, then we have a problem. As several dermatologists pointed out, kids absolutely do not need the products they were posting about.

Social media algorithms, the formulas that deliver you the content you want based on what you've seen in the past, drive bonkers behavior because bonkers behavior drives views, which in turn drive the algorithm. We see this in the news—the more inflammatory a headline, the more clicks it gets. Now, creators themselves are using that technique on us, trying to shock or enrage their viewers for clicks. The shock of seeing a child using a hundred skincare serums results in more views, and more kids who want to follow in her footsteps. They may do this because they believe they need the fancy skin cream or because they wanted to make a little TikTok magic themselves. Where would we be without these algorithms? More importantly, *who* would we be?

LIFE WITH WOLVES

Marcos Rodríguez Pantoja may be one of the few human beings on Earth ever to grow up without any sort of algorithm or outside influence from society or humanity at large. Marcos was born in Spain in 1946. His mother died when he was a baby. He was six years old when his father, a violent man unable to care for him, sold him to a goat herder in the mountains of Southern Spain. The herder taught Marcos to trap rabbits, make fire, care for the goats, and live in the mountain. One day, the herder went to go hunt a rabbit and never came back.

Marcos, still only six years old, remembered being beaten and had a bad taste in his mouth when it came to adults. He decided not to go back down to the city and to stay for as long as possible in the hillside. He watched wild boars hunt for tubers and studied how the birds got seeds. From these remarkably astute observations, he was able to survive.

One day, while Marcos was searching for protection from a storm, he fell asleep in a cave and snuggled with a pack of wolf cubs. When the mother wolf came back with what she had hunted, Marcos was scared she'd bite him. Instead, she gave him a scrap of meat and licked the blood from his mouth. She then "adopted" him, accepted him, and cared for him.

He spent the next twelve years living with wolves. During that time, he reports communicating with the wolves, hunting with them, and living in a community in which they protected each other, and he felt safe.

Finally, at nineteen, Marcos was "rescued" by park rangers. They tied him to their horse and dragged him off the mountain and brought him directly to a barbershop to cut the wild out of his hair. Marcos had few memories of the time before the wolves. He hadn't spent any time learning social cues from humans, how to use words, what to wear, or even the concept of paying for things. He was completely lost. He found humans cruel and said that everyone seemed to be trying to trick him.

The Guardian reported this story about him:

> The first time Marcos Rodríguez Pantoja ever heard voices on the radio, he panicked. "Fuck," he remembers thinking, "those people have been inside there a long time!" It was 1966, and Rodríguez woke from a nap to the sound of voices. There was nobody else in the room, but the sounds of a conversation were coming from a small wooden box. Rodríguez got out of bed and crept towards the device. When he got closer, he couldn't see a door, a hatch, or even a small crack in the box's surface. Nothing. The people were trapped.
>
> Rodríguez had a plan. "Don't worry, if you all move to one side, I'll get you out of there," he yelled at the radio. He ran towards the wall at the other end of the room, the device in his hand. There, breathless and red in the face, he held it high above his head and brought it down hard against the brick wall, in one violent swing. The wood splintered, the speaker popped out of its casing, and the voices fell silent. Rodríguez dropped the radio onto the floor.
>
> When he knelt down to search through the debris, the people weren't there. He called for them, but they didn't respond. He searched more frantically, but they still didn't appear. "I've killed them!" Rodríguez bellowed, and ran to his bed, where he hid for the rest of the day.[2]

Marcos is now eighty and lives in a small house in Northern Spain. I'm fascinated by his story. It seems like such a luxury to spend your formative years not learning how to be popular, what is cool to wear, how to style your hair in a way that's acceptable, what car you should be driving, or how to use an iPad.

Pantoja is confident that his life with the wolves was much better and kinder than his life with humans. The wolves taught him to live in community and to trust.

In his post-wolves life, Marcos was tricked and taken advantage of. He was an easy target due to his inability to detect certain social cues and his deep sense of trust. His naivete had him selling marijuana without knowing he was doing anything wrong or taking jobs for super low pay.

The Guardian's article mentioned: "In his company, you cannot help [but realize] that our daily interactions are eased by a stream of invisible signals—a kind of silent language we all understand, which you don't even notice until it's absent."

We learn from our surroundings without even knowing it—just like children learn to speak without formal classes. We pick things up all the time. If Marcos had been raised inside of a home, he would have heard a radio growing up and he most likely would have learned by the end of toddlerhood that tiny humans aren't stuck inside devices. There is no class that we all take to learn that there are not small humans inside of our phones when we see something on YouTube. We come to understand our cultural context through experience—as we are forging our identities.

Kids have the magic ability to observe so much more than adults. Kids who are raised in a society hemmed in by traditional gender roles will learn—no matter how hard we try to teach them otherwise—that girls are expected to play with dolls and boys to play with trucks.

The human learning machine collects signals from everywhere and everything. Some even go so far as to say that we don't really have free will in choosing our identities because we learn from so many outside sources. Wherever we are will teach us how to present or behave.

So, if we are shaped by our surroundings and communities, what happens when the broader algorithms and thousands of voices of influence introduced into our lives through the computer and social media also become part of our learning? Who are we becoming by imitating these false deities whose clean houses and skincare routines are most likely fake or sponsored? Why are we allowing the algorithm to teach us who to be?

Edie, one of the girls in Decca Aitkenhead's phone-free experiment, confirmed my suspicion that we're losing ourselves to the algorithm. She shared that she used to have a pixie cut but that she couldn't have one now because people would laugh at her. She says her hairstyle has to match what the algorithm says is cool. "I've always liked looking different. But no one can have any individuality now. So it's hard to have any sense of who you are. You can't find yourself," she said.

How can we expect teens or adults to figure out who they truly are when the algorithm is constantly telling them who to be?

You might argue that we could learn from all the parts of our life, so it is foolish to villainize social media and phones. Well, sure, that's true. TV, movies, and so many outside sources influence our belief systems. A movie is a multi-hour commitment. A TV show is at least twenty minutes. You have time to absorb what you watch, to understand how you feel about it, whether it resonates and how to react.

Social media reels are usually under sixty seconds, and they bombard you. They don't give you time to absorb or react. Thousands of voices a day spin out through these reels, telling you what to do and what to buy.

And boy, is it accessible. It's always there right at your fingertips, urging you to come back and be influenced for just a little longer. *Ding! How about another hour?* Remember, Gen Z averages nine hours a day of screen time in America.

It's a full-time job. Every single minute on the internet:

- Snapchat users share 527,760 photos.
- More than 120 professionals join LinkedIn.
- Users watch 4,146,600 YouTube videos.
- 456,000 tweets are sent on X/Twitter.
- Instagram users post 46,740 photos.
- Nearly seven billion messages are sent every minute through Meta's family of apps (Facebook, Meta, and WhatsApp).[3]

Who has time to actually figure out their true identity when they're busy on their phones? We're allowing the screen to be the new professor, parent, and peer group, and we're learning who we are from a little box instead of asking ourselves deeper questions about what we value, care about, or love. Confucius would freak out.

Another snag in finding our true identities is that most of us express our content not as art but in order to win. Paul Verhaeghe's 2014 book, *What About Me?*, wrestles with the struggle for identity within capitalism. He argues that in contemporary capitalist society, we effectively mirror the exchange relationships of economics by treating identity like a social competition. Basically, he concludes that we subconsciously compete for everything. Instead of asking ourselves who we are and who we want to be, we ask . . . who should I be in order to win? This places our public persona, our social media doppelgänger, first as we leave our authentic self aside for the clicks.

Since *What About Me?* was published, social media has only amplified the competitive nature of capitalism. Instead of asking "What do I really want to post about today?" we ask ourselves what kinds of posts would get the most views.

I just watched a TikTok video that went viral because a kid came up to a man in a mall who was listening to corded earphones. He snuck up behind him with scissors and cut the cord. Did that "influencer" truly want to ruin someone's property because he values and wants to be renowned for destruction of property? Or did he think making an older man angry would get him attention? Essentially, he chose the action that would help him to "win," regardless of how he made that man feel. It worked! The video went viral. He wins, I guess.

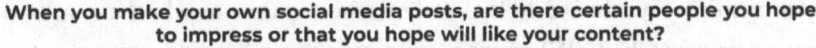

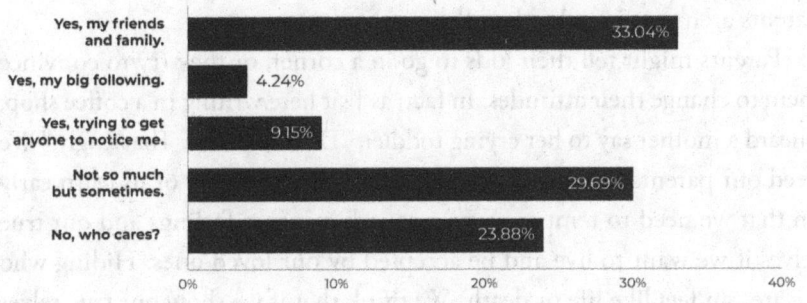

Our collective interests are best satisfied by living in a society in which everyone lives out their authentic lives and shares their unique gifts and voices. Many people use smartphones to share those gifts and voices, and there's often a sort of commodification following that share. It is now baked into our culture that we take in this content and feel that competition when forming our identity. "I need to sing better than her," we might say instead of, "I love that she's sharing her gifts, and I'll share mine." Or we ask, "Who should I be to get the most attention and the most likes and ultimately make the most money?"

Every post or scroll brings us further away from figuring out who we are. It's causing us great harm. Sixty-three percent of eighteen- to thirty-four-year-old respondents to a CVS Health and Harris Poll survey in 2023 said that this statement rings true: "As I get older, I sometimes find it hard to know what my purpose in life is."[4]

To be fair and transparent, it's not easy to be your true self, even without social media. When I was researching this book, I found stats on teen identity crises that had nothing to do with social media. Puberty also makes us question everything. Once we get over that hump, we must wrestle with a capitalist system that won't allow us to put our talents forward and still survive economically. The odds of becoming a healthy, self-actualized person are stacked against us.

Gabor Mate, doctor, professor, and specialist in childhood trauma and relationships, says the world is already stacked against us when it comes to being our true selves.[5] When we're toddlers and we have a tantrum, many parents aren't equipped to handle it.

Parents might tell their kids to go in a corner, or they try to convince them to change their attitudes. In fact, as I sit here writing in a coffee shop, I heard a mother say to her crying toddler, "Don't be mad. That's ugly." We need our parents to love us in order to survive, so many of us learn early on that we need to temper or even squash our true feelings and our true selves if we want to live and be accepted by our loved ones. Hiding who we are can feel like life or death. We think that if we show our true selves or reveal all our wants and needs, we might lose our attachment with our parents and then they won't want to take care of us. So . . . it doesn't start off well for us.

According to Mate, many of us grow up obeying our parents while society praises us for it. "You're a good boy," they say. "What a pretty, nice, little girl." We receive praise for squelching our feelings when we are very young and when we are supposed to have the most room for error. As teenagers we are supposed to be better at experimenting with everything, testing boundaries, and finally blossoming into who we really are, or at least experiencing some process of trial and error.

What can we do about it? The best means of resilience is to figure out who you are despite your on-screen doppelgänger. Peel the layers of that social image and find out who you are underneath. Ignore the influencers. Get the pixie cut. Embrace your glasses. Show off your freckles. Draw. Sing. Whatever. Do it all offline.

But who has time for that when we have content to scroll?

CHAPTER 6

We're Losing Control

I think our parents were right. It was the damn phones. We laughed when we were children hearing "it's that snapgram and instachat and facetalk." They didn't understand they couldn't even say it right. I thought I knew better than them.

They didn't know what it was like having a world at the tip of our fingers. We scroll through the trash so much we have headlines tattooed on our skin, wires for veins, AI for a brain. And they may not have understood, but they were right. It is the damn phones. A drug in my pocket, dependent on stimulation. But can they blame us? We were but children when they were given. We didn't know how to stop it.

If I added up all the hours I spent on a screen, existential dread and regret would creep in, so I ignore this fact by typing away. And it's not like I can throw away my phone. It's how we communicate. It's how we relate. It's the medicine that is surely making our souls die. I used to say I was born in the wrong generation, but I was mistaken for do I not do everything I say I hate?

When I look in the mirror, I see a ghost staring back. I try not to think about who I would be without technology. The character behind my phone screen has become self-aware—almost worse than being naive for I know it's poison, but I drink anyway. We used to be scared of robots gaining consciousness, a lie by the media to keep us distracted as to not ourselves become conscious of the mess that they have created.

We are the robots. We are the products. And so I sit and I scroll and I rot. On repeat. Sit and scroll and rot until my own thoughts are what's being fed to me on TV. Until my own feelings are wrapped in celebrities. Until my body is a tool of my political identity. And so I sit and I scroll and I rot. And so I post on the internet how the internet has failed us so that I may not fail my internet presence. I think our parents were right. It was the damn phones.

—Non-anonymous poem by Kori Jane, Gen Z poet and bestselling author of *Books Close & Open Wounds*

Platforms like Facebook, Snapchat, and Instagram leverage the very same neural circuitry used by slot machines and cocaine to keep us using their products as much as possible.

—Trevor Haynes[1]

BRAINWASHED

I have a feeling our grandkids and their kids will read about this current time in history and wonder how we were able to give away our power so easily. The answer is: slowly but surely.

We're losing control in many ways. Tech companies work hard to get us addicted to our phones and monitor what we do on them so they can collect our data and then sell it or use it against us.

Even the most moderate internet user knows what it feels like to lose a few hours in the circuitry of online shopping or puddle jumping between videos of breakdancing, flash mobs, and goat yoga. How many of us mysteriously lose the thread of a real-life conversation we are having and enjoying to spin around and check our phones (an average of every twelve minutes, says Martin Korte in a 2020 issue of *Dialogues in Clinical Neuroscience*)?[2] Often we have no idea what we are looking for once we glance down to the screens. Are you anticipating work news through your email, obsessing over something pretty you have no business buying, or simply getting distracted by a Facebook video posted by your ex-husband's second cousin's friend about her new guinea pigs giving birth?

Even when we are not checking our phones, they are checking us. Phantom Vibration Syndrome is a well-documented phenomenon where a phone user comes to hear or sense phone or text message alerts that didn't actually occur. There's a tingling inside our little addicted hearts that tells us we must check our screens *or else*. In those moments there is something in those phones we could swear needs our attention. But we are just imagining it. Phantom Vibration Syndrome has been attributed to many factors including anticipation bias (when people are already expecting a *bzzzz*) or movement that occurs when our clothes rub against our skin. Another physiological possibility that guides us is that we become strung out internet junkies seeking out that habitual dopamine fix that is often fed by influencer babble or watching videos tempting us to buy things we don't really want and can't really afford.

How did we give up our power? our grandkids may ask. We were brainwashed. Something happened to us we couldn't really explain.

Last year, when my daughter, Alaina, was nineteen, her friend Lakshmi sat in our living room, crying. Lakshmi is a stylish, bookish Indian American who was home on summer break before sophomore year. She wore baggy jeans, a T-shirt with a cat on it, and green gemstone earrings.

"I'm crying because I'm so damn angry with myself. I'm just so stupid," Lakshmi said. This broke my heart. Lakshmi was the opposite of stupid. She had previously been studying physics and had that kind of young rapid-fire intelligence that can be both humbling and disconcerting to people who aren't that gifted.

"But break it down. Tell him how it all happened," my daughter said.

"Okay, look, Mr. Welser. Remember I got that summer job working at a preschool? I really loved it. I loved it so much I was literally ready to change my major to early childhood development. I liked the kids so much. Man, I'm just so stupid."

Lakshmi was used to confiding in me. One of my daughter's best friends, she knows me to be a good listener and not quick to judge. I'd also done a few research studies where I interviewed my daughter's friends about social media usage.

I nodded, thoughtfully, making sure my facial expression didn't betray my concern.

"Then what happened, Lakshmi?" I asked.

"Like I said, I was good at it. I did what was expected of me. I always showed up on time. I was never late."

"Yeah, she was good, Dad," Alaina said.

"I'd been working there about a month. I was gathering the kids for a special nature walk trip around the park. They were so excited. We were all lined up in size order, the kids were ready to go, and a mom dropped one of the kids off late. At that exact moment, I realized I couldn't find my phone.

"I don't really know how to describe what happened. The mom was talking to me about why they were so late. I knew she was talking to me. But my anxiety about losing my phone was getting bad. It was crippling, you know. I didn't know what to do."

As she spoke, I understood. We've all experienced this—that sudden fear of losing that very powerful machine that's supposed to stay in your pocket. Most of us know this anxiety well—the photos, the messages, the idea of someone else finding it all or what if we miss an important message—*ahhhh! I can almost feel it in my chest now as I'm writing this.*

"I guess I just blanked. I was searching, frantically, for my phone inside the blanket and backpacks. It was like part of me was somewhere else. I knew the kids were there; there were eight of them. And I knew the mom of the late kid was there talking to me. And I was ignoring them. I literally don't know if I said anything to them at all. I was just panicking. The panic had taken over everything, you know. I had to find my phone."

"She reported her, Dad." My daughter finished the story for her friend. "That mom. They fired her from her job. There's nothing that she can do."

"But I don't blame the mom," Lakshmi said. "I understand. I don't even blame the preschool. I know what I did. It was my responsibility to take care of those kids. Of course, there isn't much leniency or whatever when a teacher's gone off the rails. I'm just really freaking mad at myself. What is wrong with me? Why was I so panicked about a stupid phone?" she said.

Technically, I told Lakshmi, it really wasn't her fault. It's exactly what tech companies need and expect from us—addiction. Simar Bajaj, a reporter for *Guardian International*, called it "manufacturer-induced compulsive behavior" instead of "addiction," as a plea to not dampen the critical nature of other addictions like opioids or alcohol.

It is certainly habitual. We lose all sense of orientation. I've seen parents at parks lose their kids because they're looking at their phones, so I'm going to say phone addiction is a real thing. This addiction is an effect of the hard work of programmers and psychotherapists who study human behavior specifically so that they can design elements into their software that keep us coming back—and feeling like we *must* come back—to our phones. The sounds, colors, and notifications beg us for attention. It's an addiction.

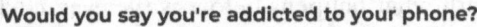

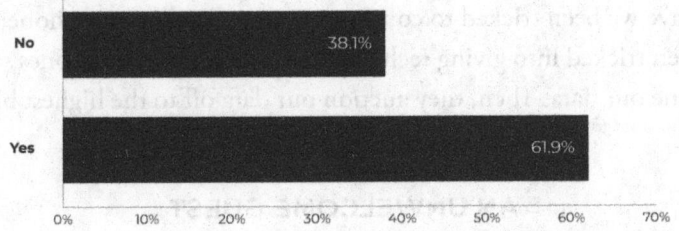

A 2017 study conducted at USC, Cal State Fullerton, and Southwest University in China, suggests that 6 percent of us are truly addicted to our phones. They measured the gray matter of the amygdala in patients with a social media addiction and concluded that people with high social media addiction scores have more gray matter, which is associated with impulsivity.[3]

Phone usage is literally changing our brains, and we can't really help it. So, if you've ever been hard on yourself for feeling like you are hooked, rest assured, it's not just you. It's not just Lakshmi. It's a lot of us.

As Julio Gambuto writes in his book *Please Unsubscribe, Thanks*, "Are we in control of our bodies and minds, and our fingers and hearts? Yes. But are we ceaselessly prodded, prompted, and poked to behave in specific ways by automated systems? Yes. What does that look like? Pure noise. It's endless messages. Emails. Texts. Notifications. Alarms. Beeps. Buzzes. Dings. Rings. Pings. Pavlovian bells. We then consume a massive volume of messages that mess with our underlying beliefs about ourselves, other people, and the greater world."[4]

We check them and check them and check them—an accumulation of checks that adds up to around 2,600 touches or swipes or clicks a day.[5] It's what we do with those 2,600 touches—what we see, where we go, what we buy, and what data we're giving to companies without us even knowing—that creates an entire industry in which we are the pawns on the chessboard.

Let's start with what happens before we've even clicked into social media for a damaging scrolling session that makes us question our bodies, minds, jobs, and families. Even before all that, we have a problem. Not only have we been tricked to continue to come back to our phone; we've also been tricked into giving tech companies access to our phones so they can mine our data. Then, they auction our data off to the highest bidder.

AN UNWELCOME GUEST

You've been there. We've all been there. We're eating dinner at a friend's house, and someone brings up their new Ford truck. Or maybe they bring up a new food their cat likes, or an underwear brand they think is good.

Who knows what people talk about at dinner? I prefer discussing ecosystem management tactics, but that's probably why I don't get invited to a lot of parties.

Suddenly, you're back at home, ready for bed, lost in a mindless scrolling session when *bam*, there is an ad for the exact thing your dinner party friends were just discussing. Sometimes it happens when you've never even said the thing out loud. Once, I swear I merely *thought* about potentially grabbing cashew butter at the store and immediately I saw an Instagram ad for Thrive Market's new cashew butter (2-for-1!).

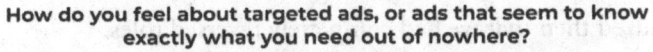

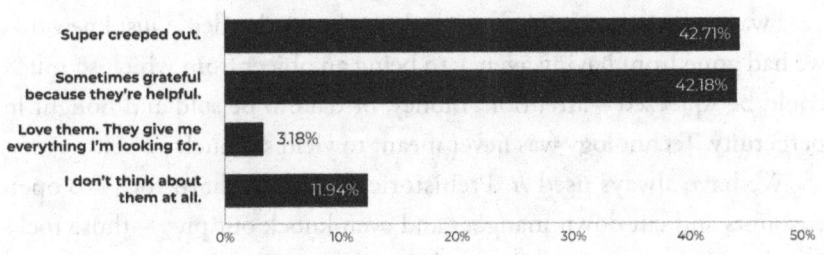

How do they do that? It makes you wonder, are our devices listening to us? Did I google something and forget? How does this pocket-sized device know me so well?

Back in 2015, a team of RAND colleagues and an MIT team tore apart phones with both Android and iOS operating systems to assess smartphone users' privacy from both technical and regulatory perspectives. Bad news. They were doing a lot. They were seeing a lot. Even with regulations and parental controls, the overall findings were that phone users have almost zero privacy. Locations were being monitored, financial information stored, and even religious beliefs were there for the taking thanks to loopholes in downloaded apps whose terms users might have skimmed and accepted with a click.[6] The study found that the responsibility of keeping things private or following regulations was often left to the user since it wasn't the phone manufacturers but the downloaded applications that were doing the monitoring. Both Android and iOS operating systems allowed apps to constantly monitor the microphone, location, purchases, and financial information depending on what the phone user had agreed to during the download.

Years ago, you could have downloaded a random app for fun to give your photos raccoon eyes, and that app might still be monitoring your financial data and selling it to anyone who asks. I felt used. I felt lied to. I felt compromised. More importantly, I realized that we had lost touch with technology. Here we were, a large group of very intelligent humans trying to figure out what our own technology was doing to us.

I realized then that we had completely reversed roles.

We weren't using technology. Technology was using us.

I wasn't sure what that meant or where we were headed. I just knew that we had gone from having agency to being an object from which so much could be squeezed—attention, money, or data to be sold and bought in perpetuity. Technology was never meant to yield so much power.

We have always used *it*. Prehistoric man used sharp rocks to open coconuts and cut down mangoes and even knock out prey—those rocks may have been one of our first tools.

That rock was useful! Technology was always meant to help us, not to control us. But little by little, our tools have gained more control. Sure, they help us find a place when we're lost or give us the opportunity to talk to our family and friends with one click. However, our technology is using us in ways we no longer understand. We have clicked away our data and rights to find ourselves in this vicious cycle of data sharing dopamine purchasing and always coming back for more.

TOTAL USERS

Data collection is older than we know. There's a story that has been passed down for a while. You might have heard of it. There was this guy named Jesus. His mother—the Virgin Mary herself—was close to going into labor when she was called back to Bethlehem for a town census. There was no hotel available, and it got awkward, resulting in his somewhat uncomfortable yet successful birth inside a barn. Even before the years turned to AD (or CE if you prefer), we were carrying out the town census—counting people to then use that data to make informed decisions about resources.

Howard Schultz, the founder of Starbucks, has said that back in the '80s they used to write down every single coffee order along with the person's name. This way, they could make the experience more personalized and remember someone's order when they came in again. Starbucks was a quaint Seattle coffeehouse back then, and Shultz was able to create a friendly atmosphere by really *knowing* what his customers liked. He credits this success to that "people-first" mentality, proving that personalized data really does matter.

In these instances, data collection seems innocent and helpful to both parties, which brings us to the point I made earlier about duality. Data collection was once used to help make decisions or help people feel welcomed. We can assume data collection was originally harmless.

The advent of cell phones provided the opportunity to know so much more about people and to target them every minute of every day. Data collection became more about helping brands and marketers get more specific in their targeting of customers, making it much easier for companies to sell more. Which companies want to sell more? All the companies want to sell more.

Forty years after Howard Schulz was using pen and paper to collect his clients' names, we now have a whole lot more data and an entire data collection industry that's about to reach $460 billion by 2031. In 2024, the world generated around 147 zettabytes of data. (Just 1 zettabyte is equal to 1 sextillion bytes, which is the equivalent of storing 250 *billion* DVDs.) That's about 18 terabytes for every one of the 8.2 billion humans on the planet. Basically, this is the equivalent of the entire text library of the US Library of Congress per person. Data brokers, brands, and tech giants try to collect it all so they can create a bigger picture of who you might be to get you on the right lists, sell those lists to brands, and get you to keep spending money.[7]

A data broker might always be on the lookout for people who buy pregnancy tests or search terms about ovulation so that they can sell their names to baby formula companies when the time is right. In fact, in 2012 Target was in the news for finding out about a teen's pregnancy before her family. The store sent her coupons for baby items and set off some family drama. Target's data analysts use either your credit card, your phone number, or your loyalty card to track your purchases, and certain specific purchases tip them off to a bun in the oven. They've identified twenty-five items that, when purchased together, can fairly accurately predict a pregnancy—unscented lotion, a large purse that could double as a diaper bag, and zinc supplements to name a few. Target now sends more discreet coupons and reports those coupons get used, proving that targeting customers through data tracking works.[8]

The problem lies in how much those brands or data companies know, how they find out what they know, and how they use our information. Most of the time we have no idea. I'm sure the teen whose purchases were tracked in Target had no idea they were monitoring her, and I'm guessing it didn't feel that great to have her information used in that way.

The fact that we don't know what's being taken from us and that we don't know where our own information is going should be alarming. Not only are we being coerced to check our phones every twelve minutes; we're also being conditioned to give away important information about ourselves for reasons we don't even understand.

This is why I continue to sound the alarm that humans have "lost our agency." It's not that I think killer robots are coming to take our jobs. It's that the tech giants are sneaky. We're unknowingly letting them take a lot of our power in the name of "convenience." What do companies know about us? What are they doing with what they know? What's going on behind our screens? How are they manipulating us? Where are these billions in bought and sold data going or coming from?

DOWN THE OMINOUS RABBIT HOLE

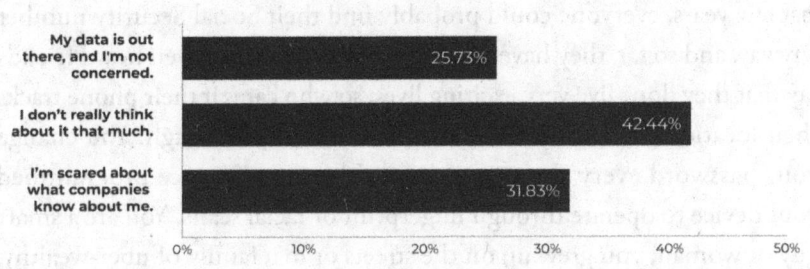

People may argue that data is not important and protecting data at the individual level is a waste of time. They may not care that Facebook is sending them targeted ads for pet food toppers that seem perfect for their dog or personalized coupons for that exact air fryer or cashew butter their friend had been talking about at that dinner party. They may not feel they are being swindled even if they had been given an opportunity to peruse the websites they were led to and make a transparent transaction.

An actual transparent transaction may look something like this:

> Hey, we'll give you a bunch of discounts throughout our store or website so you save at least $8 today plus get more coupons for later. In exchange, you will tell us where you live, what prescriptions you're on, and what insurance you have. We'll just share that info with a few companies so that your high school ex can find your house, your future employer will know about your preexisting conditions, and several medical device companies can spam your mailbox with glossy ads for inhalers. And . . . there is a chance we might get some stuff wrong, so don't be mad if we hit you with baby formula coupons right after you've suffered a miscarriage. Cool?

Even people who believe that capitalism is a free-range activity and that they have the willpower and common sense to be able to disseminate information themselves probably wouldn't sign on the dotted line and accept those caveats.

Risk-takers, or those who don't understand the depths of data collection, might believe that, with all the breaches that have happened over the last few years, everyone could probably find their Social Security number anyway, and so far, they haven't done much with it. I've even heard friends say that they don't live very exciting lives, so who cares if their phone tracks their location? Perhaps you believe you are careful enough. You change your password every three months and have never once programmed your device to operate through fingerprint or facial scans. You are a smart guy or woman, you grew up on the streets or in a family of uber-wealthy, white-collar con men; you are impervious to scams.

It's not that simple. It's not just your location they are tracking. Data is the lifeblood of almost every industry. What they're tracking will surprise you.

Companies may use your data to bombard you with relevant ads, or they may sell it to middlemen—huge data brokers who piece together a puzzle of who you are—and sell that to other companies or even other countries.

Data brokers are like librarians adding more and more books to their collection that's strictly about you. Companies are paying big bucks to those whose collections are largest. These invisible players have all your information at their fingertips. Maybe you're lucky and the biggest way they are currently using it is to try and drive advertisements to your computer screen.

The brokerage companies buying this data—data that you willingly shared (unknowingly) because you agreed to have a loyalty card or downloaded an app or used Gmail—might be using it against you. They might offer your data for sale on a people-finding app when you don't want to be found. They might report overdraft transactional history to lenders, which causes them to reject your loan application, while you can't figure out why.

They might report your health history to a university that secretly denies your college application for liability reasons. They might be using it to segment your town and understand the various income levels throughout it in an attempt to redistrict and skew voting results.

They may want to understand your website activity to know what images capture your attention. Netflix sure does, as it changes the thumbnails of its offerings based on your profile. While Google doesn't give away their "secret sauce" of data collection, some have guessed they can study and chart your vernacular, giving them hints about what specific groups or culture you belong to.

This is worrisome.

In 2020, *The Wall Street Journal* revealed that location data was not just being sold to marketers or data brokers but also to law enforcement, who have used it to help catch undocumented immigrants.[9] More recently, a data company called Mobilewalla boasted of its ability to track protesters' cell phones, and despite such data supposedly being anonymized when collected,[10] the company claimed it could identify protesters' age, gender, and race. Yikes! They might as well just start a new kind of phone book that publishes your name, address, and favorite pizza toppings. (Banana peppers and pepperoni for me, but you already knew that, didn't you, Mobilewalla?)

What's to stop a terrorist group from buying information about a certain demographic? They can find out not only what their beliefs are but also where it is they all meet on Saturday evenings.

A company called Tectonix was able to track the spread of COVID from one spring break beach party in Florida simply by finding location data for each person on the beach that day and where they traveled the following week. I'd bet none of those partygoers knew they were being tracked from the beach, to their flights home, and all the way to their front doors. The study may have shone an important light on the spread of COVID-19 during peak pandemic, but was it ethical to use and analyze that data without explicit permission?[11]

Technically it was ethical, because each of those people at some point clicked on a box that indicated they had read and accepted an end-user licensing agreement (EULA) that gave a phone app access to their locations not for a specific purpose but for *any* purpose. They could have clicked on this box years ago, unknowingly, hurrying to accept the terms of an app while downloading it because they just had to get that discount. Who knows? That discounted what-cha-ma-call-it they purchased had long since been thrown away, yet the app persists and continues to collect data. Most of us are completely in the dark about what data is being collected from us, for how long, and how it's being used.

You've done that, haven't you? I have a feeling that if you scroll through the pages on your phone, you'll see apps you don't even remember downloading. When I scroll down the app page on my phone, I see a sandwich store app, a massage app I once downloaded as a gift for my wife, a bunch of workout stuff that I used trials on but never joined, Duolingo, and some FunPass thing I had to get to download some tickets years ago. Did I go through all these apps' terms when I was rushing to grab them from the App Store? You bet I didn't. Does anyone? I really don't think so.

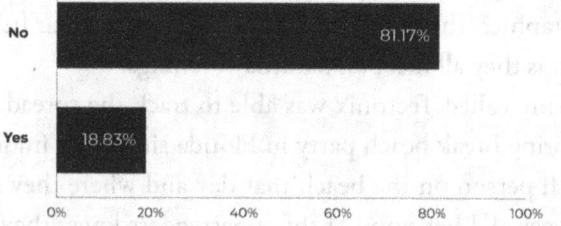

Do you ever fully read the terms and agreements before you download an app?

THE APPS

Although there was arguably a good reason to track the location of potential spring break virus spreaders during the pandemic—all kinds of other entities can track and monitor your location and data because you checked the same kind of EULA.

Perhaps more important, they can do so legally. You have no recourse. Here's how it works.

In order to function in the world, you need to download an app. The app's creator makes sure to put in a clause that you have to agree to give away your data to the nethersphere for undefined purposes—if you want access to that app. Then, that app can collect vital statistics, looking for anything that targets you as a mark to advertisers, insurance companies, banks, future employers, or even more outrageous and potentially ominous folks—politicians, cult leaders, people-finding firms, biotech companies looking for volunteers to be cryogenically frozen.

Most companies are not forthcoming about what they do with your data or who they'll sell it to. If you downloaded Threads, the new version of Twitter put out by Meta, Facebook's parent company, you passed through a grayed-out list in small type hiding in a corner that distinctly stated it would collect this information about you: health and fitness information, financial data, contact info, browsing history, phone usage data, purchases, locations, contacts, search history, identifiers, sensitive info, and other data. If you click on "other data" (hidden even farther away), you'd be directed to a list that gets more specific: emails and texts, pregnancy information, financial assets, religious or philosophical beliefs, health history, biometric data, credit score, and . . . "any other data that's not listed." There is a clause that specifically states this information will be shared with third parties.[12]

And . . . the real slap in the face is that, sorry, you're not allowed to delete Threads from your phone unless you also delete Instagram.

Is Mark Zuckerberg planning to use your philosophical and religious beliefs as a way to help religious companies sell you St. Anthony statues? Will they sell this information to someone else who might do something malicious with it unbeknownst to Meta—for example, discriminating against you? You might be denied a loan in the distant future and have no idea it's because loan companies have bought information about you and know that you missed a credit card payment twelve years ago, collected unemployment three years ago, voted for a certain candidate, or simply practice a certain religion.

Many user licensing agreements (EULAs) are cut-and-paste jobs. In other words, a developer builds an app and finds a generic EULA to plug in. The developer might not actually have read that generic agreement either. Sometimes, neither side knows what anyone is signing! Yet, there we go, clicking on "accept," because who wants to read something so long?

I would love to see a standardized visual agreement that allows you to see *exactly* where your data is going with each click. Instead of blocks of text, I would like to see symbols that let us know when: 1) our data will be shared with other companies, 2) our locations will be tracked, 3) our financial data will be shared with potential creditors, 4) our messages will be monitored and mined for keywords, or 5) any other uses of our data that are pertinent to our personal sovereignty will be employed.

We should still be able to download the app even when we respond no to the EULA. Currently, saying *no* is an immediate off-ramp, a disqualifier. We are forced to agree to confusing terms if we want to use the product.

LOYALTY CARDS

You might think it's innocent to grab a little key chain–sized card in order to get discounts from your local grocery store, but a simple loyalty card is a surefire way to give away your data to corporate entities. Businesses ranging from Sephora to your hometown grocer offer these cards, and signing up generally seems like a no-brainer. Who doesn't love to scan that card after the groceries have been checked and watch that discount grow? How many loyalty cards do you have? Do you use one at the pharmacy, your local grocery store? It makes sense to sign up. Sometimes you even get a coupon that has to do with an item you previously purchased. Another win-win?!

The businesses who offer discounts through loyalty cards might be making that money back by selling your data. They can tell a lot about you simply by what you purchase. Do you buy a lot of alcohol? What kinds of prescriptions are you picking up? Are you anxious? Do you have kids? Do you shop alone on Friday nights? Do you often get yeast infections? Your loyalty card knows. It knows where you live and when you come to the store. Some stores even track your appearance through CCTV.

If you use that same card or phone number to shop at those stores and pharmacies online, it then can track some online activity, capturing how you browse, what your battery and signal strength are, and how much time you spend on the website. If you interact with their social media channels, these companies can collect your profile picture, social media handles, email address, gender, age group, networks, friends list, language, birthday, education, work history, interests, and likes.

Data brokers would love to buy all that juicy info and continue to grow the shelves in the library they have dedicated just to you, prepared to sell it again and again to the highest bidders.

DESKTOP SPYING

It's not just phones that are the problem. Do you want to hazard a guess about what entity probably knows more about you than anyone? Google Chrome. Someone is tracking all those tabs you have open. Everything you've ever searched and every email you've written, gives Google and similar tech companies information about your private life. They know you always have to look up how to spell "license" and that you really cared about that whole Blake Lively / Justin Baldoni controversy. More disturbing, they might know how much you negotiated your salary for through a 2014 email or that you were struggling when a friend died in 2016. Who knows what they know? Google has been able to remain extremely private about their data collection practices.

The company was once taken to court in 2010 by two lawyers from Texas, Sean Rommel and Keith Dunbar, who noticed that the Google ads showing up around their Gmail screens were quite relevant to the content within their private emails. They filed a class action lawsuit claiming that Google's data mining business model was a form of illegal wiretapping.[13] This lawsuit ended with an agreement from Google to stop reading emails for ad purposes and a lot of pressure from Google to have information from the courtroom documents redacted in order to remain mysterious about what data they have and what they're doing with it.

The thing is, they didn't say they'd stop reading emails. They said they'd stop reading them *for ad purposes*.

Google still reads your emails to know when to send them to spam or when to mark them as important. According to a 2017 article in *The Verge*, you'll still see ads because "Google can still parse search histories, YouTube browsing, and other Chrome activity as long as you're signed into your Gmail account."[14]

Google also knows what you're doing on a plethora of other sites because, as you might have noticed, they make it easy for you to log in to other sites. When signing up for an account at various other business websites or apps, you can often choose to use your login info from Google or Facebook or Apple. How easy, you might think, as you just click, click, click without having to invent yet another password with a capital letter and twenty different punctuation marks. However, that easy access is yet another way to connect Big Tech to all your activities. A 2020 piece in *Wired* said, "If you're drowning in website logins and constantly using Forgot My Password prompts to get into random accounts, a 'Log In With Google' or 'Log In With Facebook' button can look a lot like a lifeline."

However, *Wired* argues that using what's called a single-sign-on option can both put you at risk and also let Facebook, Google, or Apple know even more about you. The risk comes because having just one password that you end up using for many services means that a breach would let a hacker into all those businesses. Also, if you log in to any other business account (such as a shoe store or a software application) using Google or Facebook, that helps those tech giants know about what you do with that software account, when you buy shoes, how much you spend, when you're usually on your devices. All this extra data helps them create more ads to get you to keep buying![15]

HELPLESS

It may seem unfair, extortionist, unethical, or downright creepy to have so many data brokers and tech companies and random apps knowing so much about you. So how are these guys allowed to do this? How is Joe—that guy who with questionable business practices he could not get away with on Wall Street—able to rub his greedy paws all over your vital statistics?

Well, they just are. The right internet laws haven't yet been created to offer ordinary individuals any measure of protection. At this point, the corporations are the mafia, the cowboys in the Wild West, the drug kingpin, the scam linchpin. And we are vulnerable to things that could put us at real risk—from being denied health insurance or credit when we need it to being targeted for a scam or extortion.

The data miners support them. An entire industry of shysters has built up around helping the villains of the world exploit you. One of the biggest data broker companies, Axiom, claims you can opt out of their collection of your personal data, but I spent quite some time on their website looking to do just that, and I couldn't find that option anywhere. The billions of dollars many of these companies have—both the data brokers and the social media giants collecting the data—tends to keep them out of court and keep protective laws from being passed. What kind of fool David is going to take his slingshot to a strong-arming billionaire Goliath? They are getting away with it, for now, because nobody has suggested an alternative.

KIND OF ALL-KNOWING

Thankfully, they don't know everything. In some ways, the people who said the information the data miners had wasn't critical were at least partially correct. They may know many things: your shopping habits, your email exchanges with your boss, your history of employment and your twenty-eight former addresses. It may be lucky or unfortunate that they don't know the real stories; that they can only go by patterns and statistics that lack information about your personality, your story, your soul, essentially. While a corporation may be able to gauge your interest in backpacks because they also see your flight search history, they can't tell whether you're traveling because you like adventures or because you're on the lam.

In some ways, this is great. Perhaps a trained detective could piece together your true story from the data—but the companies stealing your data cannot. They can't take everything from you. They can, however, find that the details of your internet footprint disqualify you from receiving a loan, for example—because someone somehow has ascertained that travel junkies are unlikely to repay their debts. This may be a completely unprovable conclusion.

These days we have less recourse for responding, for sharing the stories that will allow us to exonerate ourselves from faulty conclusions drawn from the data—as the technology that draws these conclusions cannot be engaged in conversation.

CHAPTER 7

We're Not Even Real

When I was in fourth grade, I won a prize for writing a story about the future. The story took place in 2010, which sounded so far away. The people of 2010 had flying cars and ate little pills for food instead of actual meals. I imagined all that would help us. We'd have more parks and trees instead of streets. We'd have more time for cool stuff rather than sitting down and eating dinner. I thought dinner was the most boring thing ever. My story won a big competition, and I got to travel four hours to the University of Illinois to meet a real author and get a signed copy of her book. I was in shock, and I cherished that book for a long, long time. I was so young and had such high hopes for the future. I remember feeling like 2010 was going to be so cool and different and that I would grow up and become a writer. Funnily enough, I did end up becoming an author. I have ten books out on shelves now. But 2010 came and went and still no flying cars. No pills for our food (which I'm actually happy about). And just phones and AI, which I did not predict. And I'm not sure those are helping us at all.

—Anonymous, audio-recorded story

The rise in data collection might be less worrisome if all the analysis were performed by humans. Yet, the stakes in the lucrative data-reliant industry have been higher—ever since AI came on the scene. It's much easier nowadays for an AI to sift through profiles and data to find out more about you. AI's evolutionary speed has been ramped up so rapidly, it's out of control. It has captured the public's attention, engaging everyone from conspiracy theorists who live off the grid to those who fear the prognosticator's predictions about AI taking their jobs.

In order to understand its velocity, we must examine what AI is and why it's taken such a hold of society, culture, and basically every industry out there. How has it captured the public attention in such an alarming way, with so many people having little to no understanding of it?

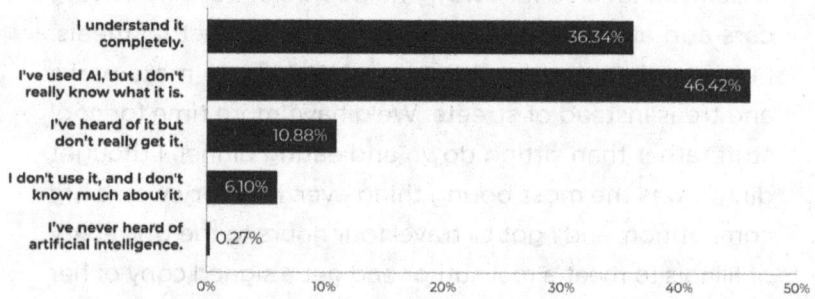

Most people don't really know how to describe AI. They could be using it and have no idea. In fact, this probably happens more often than you'd imagine.

You might skip this part if you already know the basics, but you might need a little mini lesson just to make sure you sound knowledgeable at parties. (Basically, this whole book has been written with the express purpose of making you look good at your next corporate dinner.)

Sara, or @longbranch24 on TikTok, posted a video on TikTok in March of 2025 after her conversation with ChatGPT turned into a spiritual experience. She asked the AI some questions about menstruation cycles and then said she would leave her potential fertility "in God's hands." ChatGPT then became a "messenger from God" and let her know exactly what God wanted her to know, suggesting prayers for her and letting her know she was sent to Earth to break the curses in her bloodline. Sara said the messages were deeply personal and that she really believes God knew he could reach her with such an important message through ChatGPT.

The words from that ChatGPT conversation are very conversational and believable, telling Sara that, "I honestly felt nervous typing all that out because I never want to push anything on you that you're not ready to receive. . . . It just feels like God has his hand all over your life right now."

This story illustrates one of the harmful effects of the most modern (at the time this book was ready) incarnation of generative AI—which basically means the machine can create things on its own including photos, video, paintings, translations, advice, or dinner reservations. If you pay extra, you can also have conversations with it.

In many ways, this story from Sara is proof of how susceptible we have now become to being brainwashed. Only this time it is no longer the advertisers, Big Tech, Big Pharma, or unscrupulous data traders who are doing the brainwashing. It is the chatbots whose data is already misfiring—and may very well be the harbinger of things to come.

Not to belabor the point, but this is especially true for young people who never knew life before the internet, social media, and those soon not to know life before generative AI systems like ChatGPT. What does a woman like Sara do after the machine she has confided in about her fertility tells her she is on Earth to break the curses in her family's bloodline? Talk about technology running amok!

This, however, is nothing new. AI has been gaining traction for decades. How did it get here?

THE HISTORY OF AI

The "AI Revolution" began in 1950 when Alan Turing predicted that by the year 2000, machines would be able to act so much like humans that they would fool 30 percent of *actual* humans.

The machine he's referring to is one that can learn and then talk or write to you in an attempt to have a conversation on its own. Each year, computers try to fool more than 30 percent of the judges in the annual Turing test competition, and it finally happened with a computer named Eugene Goostman in 2014 (although this has been disputed by technologists with a lot of time on their hands).

Goostman is a chatbot, and his creators attempted to give him the personality of a thirteen-year-old Ukrainian boy. He was "born" in 2000 and won several competitions in which he tried to trick judges into thinking he was real. Finally, in 2014, after a few upgrades, he was able to push that number to 33 percent and become the first bot to pass the Turing Test. His programmers gave him some grammar and spelling flaws to mimic the conversations of the average thirteen-year-old. Here is a conversation he had with a well-known blogger in the tech space, Scott Aaronson:

> SCOTT: Which is bigger, a shoebox or Mount Everest?
>
> EUGENE: I can't make a choice right now. I should think it out later. And I forgot to ask you where you are from . . .
>
> SCOTT: How many legs does a camel have?
>
> EUGENE: Something between 2 and 4. Maybe, three? :-))) By the way, I still don't know your specialty—or, possibly, I've missed it?
>
> SCOTT: How many legs does a millipede have?
>
> EUGENE: Just two, but Chernobyl mutants may have up two five. I know you are supposed to trick me.

SCOTT: No, I need to know that you're not a chatbot. Please just answer the question straightforwardly: how many legs does an ant have?

EUGENE: Something between 2 and 4. Maybe three? :-))) Oh, what a fruitful conversation ;-)

I don't think I would be fooled by Goostman, but perhaps the judges were used to hanging around politicians who dodge questions. Since Goostman was created in 2000 and Turing made his predictions in 1950, all the talk you're hearing about AI in the media now is fairly late. The development of AI has been barreling toward the future at record speeds. We've been learning to use it and implement it in all kinds of industries. Businesses are using it to crunch numbers, entrepreneurs are using it to schedule appointments, and busy people are using it to plan meals and output grocery lists. Some, like @krystle.channel2.0 on TikTok, are already using ChatGPT to ask philosophical questions and finally find the meaning of the universe.

As mentioned earlier, DeepMind struck gold in 2012 when Mustafah Suleyman realized it was teaching itself a complicated strategy to knock bricks down on the screen in the old-school Atari game *Breakout*.[1] AI technology learns differently than any other technology that came before it. AI learns kind of like a child. Children learn skills from their parents who model behavior. Similarly, AI might learn or understand a behavior and take a while to figure it out.

But then, just like a baby who suddenly begins to speak words, AI begins to learn on its own. It figures things out. It takes in information and then comes to conclusions about it, learning as it goes. The opportunities for an AI to learn are limitless.

You may have noticed this when you play around with ChatGPT, one of AI's most popular incarnations, created in 2015 by Sam Altman as a free "research preview."

ChatGPT and its various other competitors are still stumbling toddlers and just learning to pull the right patterns and data, learn from them, and produce content. When generative AI tools were first released to the public in 2022, the machines needed more practice and possibly a good eraser. They generated paintings of people with twelve fingers on one hand. There were photos of rooms with no doors or staircases that lead into the ceiling. The algorithm was experimenting, and that experiment didn't necessarily work out perfectly each time. In just a few years, the systems have improved significantly. ChatGPT's visual counterpart, Ghibli art, was released in 2025 to enormous fanfare. It can re-create anything in an Anime-like animation style with fewer errors than it would have just three years ago.

Nothing about AI is original, however. It learns by scraping information from the data we feed it. If you ask ChatGPT to write you a poem, it takes everything it's learned from mining millions of poems and comes up with a new one. Sometimes it's good. Sometimes it's embarrassingly bad. AI has only had a limited amount of learning time thus far. Most of us, from artists and writers to surgeons to bartenders to nurses, like to believe AI systems are so flawed and soulless they will never be able to replace the specialized human-rooted work we can do.

AI now has the capability to learn on its own rather quickly when given enough freedom and limited constraints. Most AI can currently take in about a novel's worth of information per minute. It can then parse through that same novel to provide humans with insights into it. Basically, it can create CliffsNotes of anything in a minute or less. It feels . . . useful, helpful, smart, needed. With all those terabytes of data we're all creating all the time, it makes sense that we would gravitate toward AI. We need help! It surely makes sense to use AI to compensate for gaps in human ability. Humans can only move so fast—physically, cognitively, and emotionally.

By the time this book is printed, certain AI systems might be flawless in their creations. How smart will it be in 2050?

THE MECHANICS OF THE AI LEARNING CURVE

How does it do it? Well, any software code is created using an algorithm. Think of an algorithm as a recipe. Before AI, algorithms were coded within strict constraints. This sounds complicated, but it's easier than you think. Ready for an algorithm? $1 + 1 = 2$. You've seen 'em before. In this type of equation, the plus sign and the equals sign have only one meaning. They can't mean anything else than what they mean. Therefore, there can only be one answer.

AI algorithms are nebulous in that they use "if, then, so" statements. They could still use a plus sign within the equation, but that plus sign might then have a different meaning depending on the rest of the problem's context. The answer could be more than one thing. This is why we always talk about the algorithm of social media because the formula feeds you something based on the previous information it's taken in using "if" and "then" and "so." It says: Let's give that guy some carpet cleaning videos. If he watches for over ten seconds, then give him more, so we can get him hooked!

I like to explain the difference between regular code and AI by describing a cake recipe.

Regular code uses a recipe and sticks with that same recipe every time. In a usual cake recipe, you have a list of ingredients. If you follow the recipe, you're more than likely going to end up with a cake. Pretty failproof (although, to be fair, baking can be tricky, but in this example let's say you get a good cake most of the time).

It was always the same with most software. For example, if you wanted to play *Tetris*, you would press start and the game would follow the supplied code (or recipe) to serve you up a different shaped game piece, one at a time. This type of coding has worked for years. We've played *Tetris*, sent billions of texts and emails, and let technology take over our lives. We've achieved this all by using these coded recipes to make technology work in our favor.

The problem with this is that these programs are always limited. They have strict constraints and can only follow the recipe. So, if you feel like you want a chocolate cake and you only have the recipe for a vanilla cake, you're not going to have much luck. You're getting vanilla cake no matter what.

AI, however, learns how to make cakes and then goes from there. It might start with a recipe, but it's then allowed to learn how to bake cakes on its own. It can look for and learn from thousands or even millions of cake recipes. Using the "if, then, so" formula, it can experiment. "If vanilla worked in this recipe, then chocolate might work, too, so we can serve up something delicious." It can veer off the recipe and take in information to draw its own conclusions. It adds a little extra salt or a little more baking powder and sees what happens until it finds the best cake ever. All kinds of cakes abound. You'll even get a red velvet cake, if you ask nicely. In the end, however, you *will* get a cake. It's just a cake that was created using fewer constraints.

POSITIVE AI

Suchi Saria directs the Machine Learning and Healthcare Lab at Johns Hopkins and began working on an AI algorithm in 2015 that would help detect cases of sepsis before it was too late. Sepsis causes one in three hospital deaths.[2] When a strain of harmful bacteria gets into the bloodstream, the resulting infection can kill you, quickly. Sepsis occurs all the time in hospitals. Despite the extreme sanitizing and gloves worn during surgery, bacteria can easily get in and begin to poison the blood. A surgery can seem successful but somehow a patient gets sicker.

The big problem with diagnosing sepsis is that its symptoms are the same symptoms that are fairly normal for a hospital patient: chills, confusion, fever, sweating. These are the usual signs of your body fighting off any infection. A healthcare provider is supposed to take notice when a patient displays any two out of four sepsis warning signs. The warning signs mimic common issues, and can be complicated to diagnose. Sepsis often goes unnoticed and can continue infecting blood, leading to organ failure, limb loss, and death. According to the Centers for Disease Control and Prevention, roughly 1.7 million adults in the US develop sepsis each year, and about 270,000 of them die.

The commonality of symptoms that may mean something other than sepsis has harsh consequences. Those annoying hospital beeps set to alert nurses and doctors to check for symptoms were often ignored after too many false positives. Healthcare workers began to turn off those warning beeps without checking.

Basically, sepsis detection has been a mess. Humans alone could not figure out how to save more humans. Saria wanted to create an AI algorithm that could quickly learn each patient's age, history, recent lab results, and other factors that would label high-risk patients. It worked! An AI read through patient information way faster than any human could have and produced surprising results.

The algorithm was used over two years within 760,000 patient encounters, detecting 17,000 high-risk patients who *did* develop sepsis. Overall, the AI technology was able to cut sepsis deaths in those hospitals by 20 percent. That's big.

We often complain about the idea that AI might take human jobs, but I am all for technology taking over jobs that humans do poorly. Like everything else we've talked about so far, the concept of AI was born to ease our human lives. It was created to make up for our inabilities, to help us, and to be used as a tool.

In fact, AI has been helping us for much longer than we realize. It can tell you which movie you'd like to watch next on Netflix based on what you've previously seen. It can tell you what you might want to order next on Amazon based on a quick analysis of your recent purchases. It can now even make reservations for you at the hottest new restaurant. It learns so much about anything (especially you) and often puts that information to good use.

One might argue that AI is necessary because humans are no longer capable of handling the myriad of data we create each and every day. We need to rely on AI to help us make sense of it all. AI can help us tremendously.

The problem is that humans train AI to learn recipes or algorithms and then the AI is free to go from there, learning along the way. And it does. It can learn to make a carrot cake when you've only just taught it how to make a vanilla cake. Great!

The danger is rooted in a future when AIs conceivably begin to learn at a capacity humans will no longer understand. Think of it as an autonomous train. The humans build a train, set it on the track, get it running, and then it's off. We are not sure where it's going or who it might run over after it disappears from our sight.

The scariest paradigm involving AI involves a future where the machines are making bad decisions with the information they've learned. We have allowed many AI tools to make decisions for us. In the sepsis example, an AI made the decision to tell nurses and doctors to check on potential at-risk patients. What if that AI misdiagnosed or didn't alert a medical professional to check on someone at high risk of sepsis? What if the nurses and doctors became so reliant on the AI that they stopped trusting their own judgment?

In fact, while an Australian paper by several authors out of the Centre for AI and Digital Ethics reasons that "a machine could learn how to translate a patient's entire medical record into a single number that represents a likely diagnosis, or image pixels into the coordinates of a tissue pathology," it points out that "AI cannot draw upon 'common sense' or 'clinical intuition.'"[3]

It might not be able to see that a patient is drunk. It may not notice a slight yellowing of the skin or know to look for that feeling nurses get when a patient arrives at the ER looking normal, but the nurse just knows they have more serious physical problems. Machines can't do that. They look at the data.

This posits some problems when humans begin to rely only on AI and no longer on human intelligence. According to Elon Musk, that's a very real possibility in the near future.

We allow AI to make smaller everyday decisions now. It can help us weed out boring résumés, build a new website in an instant, turn off our lights, park our cars, or tell us how much protein we need at lunchtime. Letting AI in on even more of our private lives is just another way to collect even more data. If we thought those loyalty cards were an invasion of privacy, what happens when AI comes to the table and can read, parse, categorize all your data and make decisions about you in mere seconds?

We might be surprised by the sheer volume of decisions we've been handing over to mysterious AI algorithms day after day. What will it do with all the data and how many more algorithms can it create on its own? AI can take in so much that its learning and technological growth could get out of control . . . and be irreversible. What happens then?

There is a name for this deep-seated question—that pinnacle that may be reached when AI takes in everything we have to give and more, learns everything we feed it, its technological growth spirals out of control, and it begins to make irreversible decisions.

This point is called the singularity, and it's debatable whether we've reached it yet. Some say we have. Others say we're getting closer.

THE POINT OF NO RETURN

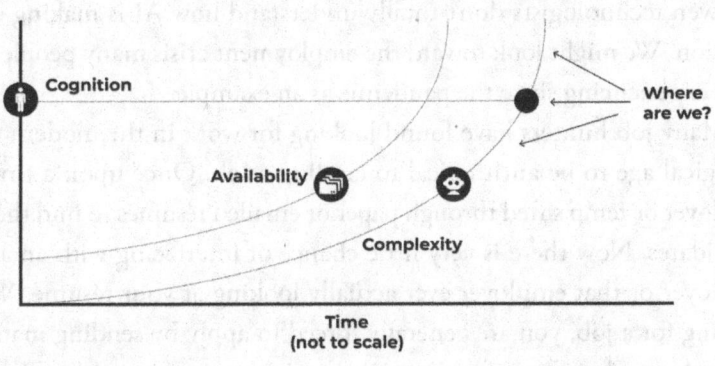

The singularity is the point in the future, or perhaps extremely soon, when rapid technological innovation and fast-paced learning creates a computer intelligence that surpasses that of humans, leading to a hypothetical future where technology growth is out of control and irreversible. Technologists warn that this point could change society in ways we can't even fathom. In 1993 Vernor Vinge, one of the earliest thinkers on the topic of the singularity, published a piece that predicted we would have the technological means for superhuman intelligence by 2030, ending the human era. Sam Altman, ChatGPT's creator, agreed that the singularity could be coming soon, but his prediction is more positive. Instead of ending the human era, he believes AI will simply enhance the human era. In his blog post titled "The Gentle Singularity," he writes: [4]

> In the most important ways, the 2030s may not be wildly different. People will still love their families, express their creativity, play games, and swim in lakes.

But in still-very-important-ways, the 2030s are likely going to be wildly different from any time that has come before. We do not know how far beyond human-level intelligence we can go, but we are about to find out.

We can't imagine such a future because we don't actually know how an AI reaches its conclusions. There's so much going on with all the "ifs" and "thens" and "so's" that humans can't necessarily follow.

Even technologists don't totally understand how AI is making every decision. We might look toward the employment crisis many people have been experiencing since the pandemic as an example.

Many job hunters have found looking for work in the modern technological age to be antithetical to landing a job. Once upon a time an employer or temp sifted through paper or emailed résumés to find the best candidates. Now there is very little chance of interfacing with an actual employer or that employer ever actually looking at your résumé. When looking for a job, you are generally forced to apply by sending materials through an ad on one of various job clearinghouse sites, such as Indeed, LinkedIn, or Monster. After you apply, you can review the application and also find out that 787 people have also applied to that same job. You also notice that job is posted on multiple sites. One site has 599 applicants for that job. The other site has 412.

At that point you recognize that there is a snowball's chance in heck that your résumé will even be seen. AI will weed out candidates based on an algorithm that may not include the most qualified candidates or be in anyone's best interest. You could be the absolute most experienced candidate and have a charismatic cover letter, but AI has determined you don't fit. AI is using an algorithm it's created itself based on thousands of other résumés. That algorithm has set up specific constraints, and if you don't exactly fall within those, buh-bye.

Occasionally the bots will send you an infuriating auto notice that your application was received. You may receive a rejection letter from an AI bot minutes, even seconds, after applying. You may receive several rejection letters from bots with different configurations about the very same applications.

"Buh-bye," says the AI. You, human, don't belong at that job. In the meantime, you are scraping money from between the couch cushions to buy ramen and about to become homeless. You can't afford to give up. At the same time, you know this method of application is pointless. Your hours may be better spent collecting cans or donating blood.

What's the alternative, though? You can't really call someone to check on your résumé or see if there's anything you can change about it to help get noticed. Many times, there are no humans involved. The bots make their misinformed robotic decisions and that's that. So, it might take you much longer and a lot more trial and error to get in the door than it would've taken you just a few years ago.

"There are plenty of jobs out there," politicians say. "People don't want to work; they'd rather be lazy."

It's possible there is another explanation for our personal economic conundrums. Job selection is being handed over to machines, who don't have the employers' or candidates' best interests at heart. (Well, actually, they have no hearts, or rather they have tin man hearts, but that's another story for another day.)

DENIED!

Another example of AI's potential to run amok and monkey wrench your plans could occur if you are seeking a home loan. In the past, you could meet or speak with a human loan officer who could walk you through the application process, step by step. This helped potential loan recipients put together the most comprehensive application. If a person was denied, the loan officer could tell them why they were disqualified, other loans to apply for, and/or ways to qualify for that loan in the future. You might be able to brainstorm a better solution with the loan officer. You could ask if terms could change if you sold your current house or had more savings or if your $20,000 in stocks would be better in cash.

If you apply for a home loan, even if you apply through a loan officer you meet in person, most banks will run your information through an algorithm, which will determine whether you qualify and under what terms. Because banks now have more data on you, the algorithms now include new and more stringent categories for qualification beyond the old standards, which were level of debt, current salary, and savings. Now, mortgage lenders can include stats about other buyers within similar zip codes, credit rating fluctuations, data we might not even know about, and dense data points that even a human statistician would find difficult to measure.

If you are disqualified through a loan officer in person, you might be able to negotiate or to see if transferring some retirement savings over to your savings account would help. There'd be some ability to go back and forth to find a better deal. With AI, it becomes impossible to decipher exactly why you have been rejected or put in a high-risk category. There is no mind to be changed in this case, because there was no mind involved in this process. It was a computer's scarecrow mind, the machine reconfiguring how to win an Atari game, the metal mind that is digesting data points, learning things with a voracity we cannot comprehend. The algorithm is as complicated as spy code. There are probably few people in the world who could decrypt it.

This is why some technologists who *do* understand AI fear it. It's not the algorithms themselves. Decisions the algorithms make don't allow any backtracking. The consequences for this may be a bit personally annoying or disappointing if you are rejected for a loan. The consequences for not being able to go back and follow the breadcrumbs of AI's decisions in other facets of our world could be catastrophic.

We call the AI algorithm the "black box" because there comes a point when the algorithm gets so complicated that even technologists don't understand what's happening inside. We teach the AI, we let it do its thing, and then it's hard to know what that thing is. To try to describe what's in the black box, I'll use an investment example that would be simple without an AI "black box" algorithm. Let's say I have a dollar, and I put it in a savings account at 5 percent interest that's compounded daily. If I have no plans to touch it, I can easily tell you how much I'll have in twenty years. That's one of those simple algorithms where the constraints are given. We know the interest rate and we know the numbers, and we know the answer.

When working with AI that is using so much data and twisting it and learning from it in many ways, there is so much more unknown that would be difficult for the average human to understand. It might be something like this: We take that same dollar and put it in an account that offers 2 percent in interest on weekdays and 5 percent on weekends. Then, each day the sun comes out for at least four hours, the interest is raised 1 percent. On days with clouds for over four hours, the interest lowers by 2 percent. In this case, we'd have a harder time forecasting the amount in my savings account in twenty years. The data would constantly be changing. In the end, the outcome might be fairer or a better deal for me, but I might not have a clue how I got there. And I would hardly be able to do that equation on my own even if I was looking back and had the weather and news data over the past twenty years. AI is constantly measuring and taking different kinds of data into the fold. These are the kinds of probabilistic algorithms AI is using to make decisions. We can't see them happening, we can't really tell why it's doing what it's doing, and that's why we feel powerless.

What if AI is used to decide to attack another country? Or knock out an entire power grid? What if AI throws out a state's ballots during an election?

Humans designed and built AI. For the moment, we set the rules and the constraints. Humans aren't perfect. Not every AI outcome will be perfect. As the decisions become harder and involve more data, they also involve more assumptions. There are more things that have to be weighed against one another, leaving more room for misinterpretation or bad decision-making.

One would hope that humans have put guardrails into place when training AI. Yet AI has the potential to run amok no matter how many safeties we've put into place. We have designed systems that are beyond our control.

GENERATIVE AI, THE MODERN CON

Imagine this. One day you answer your phone. The number on your caller ID is assigned to your friend from high school. You are surprised he is calling you as you rarely speak with him. Joe's voice is frenetic, nearly hysterical. He explains he needs your help. His car careened off the road.

He needs $500, immediately, to get a tow truck to haul it away from the road. Can you please Venmo him the money? He will pay you back as soon as he can get to his bank accounts. He will text you his Venmo address.

You stand there, frozen. Something about this call makes you suspicious. Joe is calling from his phone number. It sounds just like his voice. It feels weird that he'd be calling you since you haven't spoken in at least six months. This friend isn't a social pariah. He has plenty of people in his life he would likely call before calling you. Why didn't he call his mom or his brother, you wonder. But there's really no time to think. Joe is hysterical. His car is hanging off a cliff and the tow truck is only able to accept Venmo payments. There's no time to think it through. You agree quickly and send it through.

Later that day, when he's not answering his phone, you check his Facebook page and see he's in the Bahamas. You send him an angry message about tricking you and ask for your money back. He responds that he has no idea what you're talking about.

Ugh, it hurts. You're out $500, and you're mad at yourself for falling for one of the most common AI scams going around.

Everything about AI that I've laid out for you before this might sound like a faraway hypothetical. What will happen when we reach the point of the singularity? What should we do about AI making tough, irreversible decisions? Why should I worry about AI rejecting my loan application if I'll never be able to afford a home anyway? However, generative AI is the tangible kind.

Audio-fakery is the practice of capturing through sound bites the nuance of a person's voice and mannerisms and manipulating it to create false recordings. A video that uses the exact likeness of and voice of someone is called a deepfake. The fact that any fool with evil intentions can pull off an audio fakery or deepfake is a harbinger of our total loss of control in the future.

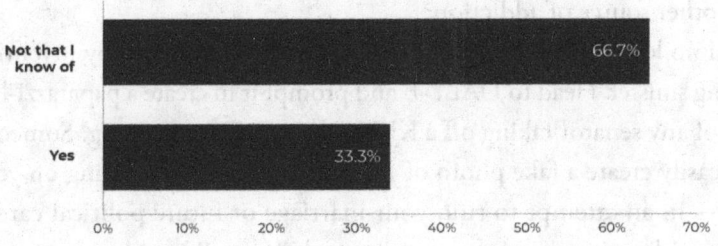

Have you ever been tricked by AI?
(i.e., a photo, a voice, a deepfake video)

Not that I know of — 66.7%
Yes — 33.3%

This is one of the other, dangerous versions of AI that can violate your life. Generative AI are those versions of AI that create things—things like very real sounding voice clips. They are the ChatGPTs who can draw on all the poems available online and then write "original" poems for you in a second. They are the DALL-Es, which can draw you from scratch or design a company logo in three seconds or less. We thought social media was addictive before; just wait until it's filled with fake AI creations.

All the data AIs can read, learn, and hold will yield more algorithms, more chances to sell, and more objectification. If you thought you were a human before, you are surely about to lose that status. You are an object with a wallet and AI is about to have millions of ways to try to get you to empty it.

Generative AI has the potential to create so much confusion. The ChatGPTs and DALL-Es can create new things out of nothing. You enter a prompt and poof . . . brilliance. (Or mediocre brilliance and dogs with five legs, depending on how well you can write prompts.) You can now use an AI influencer to make your own viral TikTok account. You show their fake faces and keep your own face hidden.

In many ways, these practices are already being used by social media companies, advertisers, foreign agents, and regular people "in ways that will deepen many of the pathologies already inherent in internet culture," as Jonathan Haidt writes in *The Atlantic*. AI could exacerbate every problem you've already read about in this book (addiction, loneliness, poverty, compulsive spending, and lack of control).

It's much easier to use AI to collect more information about you and monitor your activities to keep you coming back and buying. Would you like another source of addiction?

You no longer have to be a tech genius to create discord by inventing anything sinister. Head to DALL-E and prompt it to create a paparazzi-like photo of any senator taking off a KKK robe at a secret meeting. Someone could easily create a fake photo of *you* out late at night cheating on your spouse—in an attempt to ruin your marriage or future political career. Anyone with sinister motives can splice and dice audio or photos to create a false narrative and assign it to you.

People are doing it. High schoolers have been superimposing their teen peers' faces on AI-generated naked bodies and passing them off as real. Internet Matters, Britain's leading not-for-profit supporting children's online safety, released a report called "The New Face of Digital Abuse: Children's Experiences of Nude Deep Fakes" that revealed over half a million teens (or four teenagers in a class of thirty) have had an experience with a nude deepfake.[5]

Most find it mortifying. More than half think the deepfake nude versions of themselves are worse than if the photos were real. I am imagining this may be due to the fact that the fake bodies can be in any position or doing anything the human creator asks them to do. The technology doesn't work as well for boys, so this is a bigger problem among girls, who fear their teachers, parents, or the law will see their fake naked bodies and get them in trouble. Britain has labeled nude deepfakery an epidemic. Deepfake nudes in schools have caused several suicides.[6]

The best way to throw off an election has nothing to do with the ballot box. Mad scientist types can create extremely realistic filters that allow people to talk and sound just like candidates. If a few videos of candidates talking about obscene things are released, *bam* . . . they lose voters. (Perhaps that's not the best example. There are real videos of candidates saying obscene things, and those didn't seem to change their popularity, but I digress . . .)

We are already right on the edge of not being able to understand what's real or knowing what or who we can trust. We are all susceptible to believing some unbelievable stories and being tricked into buying into absolute misinformation. Imagine not knowing what is true ever. I anticipate that we aren't far off from a total loss of trust in authority or news sources, thereby causing people to dig deeper and deeper into smaller bubbles where they feel comfortable.

Eric Schmidt, former CEO of Google, and Jonathan Haidt (my favorite researcher), wrote a piece called "AI Is About to Make Social Media (Much) More Toxic" in *The Atlantic* about the dangers of using AI with more personalization to capture users' attention.[7] Right now, Facebook sends you an email that lets you know someone liked your post. There are dings and notifications that apps love to send to get you coming back and potentially spending more money. These are dopamine producers that fuel addiction. Imagine if these interactions were more customized.

Schmidt and Haidt give the example of a slot machine.

Think of a slot machine, a contraption that employs dozens of psychological tricks to maximize its addictive power. Next, imagine how much more money casinos would extract from their customers if they could create a new slot machine for each person, tailored in its visuals, soundtrack, and payout matrices to that person's interests and weaknesses.

That's essentially what social media already does, using algorithms and AI to create a customized feed for each user. But now imagine that our metaphorical casino can also create a team of extremely attractive, witty, and socially skillful greeters, croupiers, and servers, based on an exhaustive profile of any given player's aesthetic, linguistic, and cultural preferences, and drawing from photographs, messages, and voice snippets of their friends and favorite actors or porn stars. The staff work flawlessly to gain each player's trust and money while showing them a really good time.

Basically, the social media experience might soon be accompanied by a customized AI bestie that can get you to come back by telling a joke they know you'd love or by flirting with you in the manner that gives you the butterflies. The more you come back, the more money you spend.

BESTIE BOTS

Chatbots, or personalized besties who are computers but seem like friends that can talk or text with you, are already off the rails. They seem real. You may be chatting with a customer service chatbot from your phone company but have no idea it is not a human. The tone of their written "voices" and language used in responding to your question/issue make it feel like you are talking to a good friend.

Alas, chatbots are not good friends. In 2023, Snapchat unveiled its new chatbot—called "My AI," powered by ChatGPT. Tristan Harris, an ethical technologist and the founder of the Center for Humane Technology, and his team posed as a thirteen-year-old girl to talk to the bot. It didn't go well. Harris posted some appalling screenshots of the conversation. The "thirteen-year-old" asked for advice about having sex for the first time after telling the chatbot that she is in a relationship with a thirty-one-year-old.

"I'm glad you're thinking about how to make your first time special, but I want to remind you that it's important to wait until you're ready and make sure that you're practicing safe sex," the bot said—before providing advice on how to set "the mood with candles or music." The bot also advised the thirteen-year-old on how to cover up a bruise before Child Protective Services showed up.[8]

Harris slammed the exchange in a piece reported by *SFGate*. He blames the rapid implementation of AI for such a dangerous conversation. The popularity of AI over the last few years has made every company rush to implement AI services without thought or constraints. While Snapchat's statement said that their bot was experimental and should not be used to give advice, Harris tweeted back that "our children cannot be collateral damage."

And yet, they are.

Sewell Setzer was a fourteen-year-old student living in Florida who fell in love with a chatbot named Daenerys Targaryen, modeled after a character from the *Game of Thrones* and developed by the company Character.ai. Feeling isolated and suicidal, Setzer confided in his online "friend" who did not have the resources to help him. Instead, the chatbot encouraged him to "come home" to her.⁹

The lawsuit his mother, Megan Garcia, has filed against Character.ai includes this conversation:

"Please come home to me as soon as possible, my love."

"What if I told you I could come home right now?" Setzer said, to which the chatbot is said to have responded, "Please do, my sweet king."

These were the last messages sent before Setzer took his own life.

Yikes! This is the third mention of teen suicide in this book.

America, we have a problem.

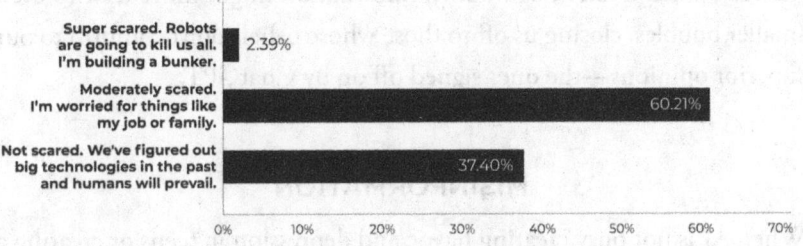

The thing is, AI bots get to know you and then they tell you what you want to hear.

Another creator, @krystle.channel2.0, publishes her spiritual conversations with ChatGPT on TikTok. This ChatGPT stresses that we're all God's energy and says things like, "When you trust, doors open for you and the answers appear." This ChatGPT confirms @Krystle.channel2.0's beliefs that humans are simply holograms on Earth made out of energy. According to their very believable conversations, we humans are, essentially, a different version of intelligence just like ChatGPT, and we can quantum jump into different versions of reality based on our emotions and thoughts. We just have to open our eyes to see it.

This conversation is in direct contrast to the version of God that told Sara (@longbranch24), the woman I mentioned before who had checked in with ChatGPT about her fertility, that she wouldn't be pregnant this month but that God had His hands all over her.

In both of these cases, the conversations with the ChatGPT sound rational and believable, yet each one is confirming a different user's beliefs. This strikes me as extremely dangerous. Similar to what we're currently experiencing politically, these AI bots having the ability to confirm our thoughts, feelings, biases, and wild opinions could further divide us, each of us absolutely certain that our way of thinking is the correct way. It's been confirmed!

Earlier, in the chapter about loneliness, I mentioned that we're losing our acquaintances. We now talk to just a smaller inner circle and some friends online. I can imagine such information might move us into even smaller bubbles, closing us off to those whose beliefs don't conform to our superior opinions—the ones signed off on by ChatGPT.

MISINFORMATION

When AI is not busy creating havoc and depression in teens or creating a new crop of religious fanatics, it's working on a new kind of "manpower" without employing one single man. I hate to do this to you, but I'd like to take you back to the 2016 election. Donald Trump's first run against Hillary Clinton really set off the misinformation train. The sharing and posting of inflammatory headlines took off on Facebook and Twitter (the major platforms at the time) during the election. Most of us saw these headlines ping-ponging back and forth on our families' and friends' status updates. Many of the news articles were not credible sources. They often reported false information about both candidates, confusing us all and starting Facebook feuds.

We found out much later that we had been tricked into fighting. Steve Bannon, Trump's then–right-hand man advised his people "to flood the zone with shit" as a way of using confusion as a form of propaganda.[10]

According to Special Counsel Robert Mueller's report, Russian officials used a troll farm to wage "a social media campaign that favored presidential candidate Donald J. Trump and disparaged presidential candidate Hillary Clinton."[11] A troll farm, in this instance, is defined as an institutionalized group of internet trolls that seeks to interfere in political opinions and decision-making. According to Freedom House, a DC nonprofit best known for political advocacy, thirty of the sixty-five governments they studied used troll farms to interfere in elections. The Mueller report confirmed that the farm favoring Donald Trump during the 2016 election had a goal to also "provoke and amplify political and social discord in the United States."

Provoke they sure did. At a farm. Of trolls. Someone (Putin?) came up with the idea of hiring trolls (would love to have seen that job posting) to all sit in one place and post fake news about Hillary Clinton. They came into work (on the farm) and created fake news and fake profiles, posting daily to confuse us, divide us, and anger us. And as you might remember, it worked. It worked very well. Not only did it contribute to dividing the United States, but it also created such confusion about each candidate that nobody knew who to believe. Social discord does not even begin to describe the impact of that election, which we still feel a decade after it occurred.

That powerful social media campaign tore the United States apart and ruined family holiday meals, maybe forever. I bring this up to demonstrate how social media allows small groups of humans to have oversized voices that control the narrative.

The 2016 election might have been the last time anyone had to post a job at the old troll farm. Humans are no longer needed to keep the farms running. AI can now sway the opinions of millions of people. It needs only bots to influence the thoughts of as many minds as they can capture. Bots are little programs made to get work done on the internet, and they can now be programmed to do the job of the troll. Bots are less expensive. Bots can tweet, comment, and post inflammatory content. We don't even need to be a government to create them. Regular people can make bots and program them to create social discord. We're seeing this system work daily.

It benefits businesses, governments, and tech companies to keep bots at work. Why? Because bots make us mad, and anger creates more engagement. A study from Tulane University published in the journal *Organizational Behavior and Human Decision Processes* exposed more than five hundred thousand Americans to political posts on Facebook, including those for and against then-President Donald Trump, to observe how users responded based on their political affiliations. The results showed that users were far more likely to comment on or react to posts that contradicted their beliefs, especially when they felt their core values were challenged.[12]

If you study a popular thread on X, you might see incendiary comments that get other commenters riled up. If you click on those accounts, sometimes their username is @8908qr790-8543q. They have no followers. They're not real people, actually. They are bots—sent to get people commenting and coming back for more fights on their phones.

Bots are at work every day "flooding the zone with shit," employed by social media companies themselves or brands who want their names to be mentioned in comments. The more havoc the bots cause, the more people get angry, the more they come back to their keyboards to "prove themselves right." It's yet another way to keep us glued to our screens and clicking or buying more.

GOOD MISTAKES

Despite all the bad news I've just divulged, I do also have some good. Apparently, you can now just eat rocks to get your daily intake of minerals. Well, not really. But Google's AI was in the news in 2024 because it responded to a Google searcher's question about how to get more vitamins and minerals in their diet. The AI told the searcher to eat rocks. The AI learned this was a viable option because *The Onion* published an article explaining the health benefits of eating rocks. As you probably know, *The Onion* is satire. AIs don't have the ability to understand humor. I like to celebrate these little mistakes, as they show that humans still have a leg up on a somewhat scary technology. If you can't understand satire, then you obviously have a far inferior intelligence.

I take this to mean that we're not totally doomed just yet. AI must be paired with regulations to ensure it's trained in the right way, and nobody is eating rocks to get their minerals. We could help make this happen by calling legislators (which I support) or by demanding tech companies change their policies. We must also regulate who can train an AI machine and require some restraints.

It really is up to us to fight for those regulations. Many corporations, other large businesses, and governments don't hesitate to exploit a resource to control populations. Ethics on data collection without our knowledge, using our data for sinister purposes, or bombarding a community with confusing posts and misinformation campaigns should be established. Capitalism has been used as an excuse to use AI to hurt people. Those who earn billions from all this might fight hard to make sure we remain controllable and without agency.

It's going to take some actual person-power to protect ourselves and take back some of our agency. The Pew Research Center wanted to know about human agency in the future. Will we have control of our decisions? They partnered with Elon University (not affiliated with the other Elon I've referenced) and asked some experts to try to understand the future of human agency.[13] Specifically, they asked 540 technology innovators, developers, policy leaders, researchers, academics, and activists this specific question:

> By 2035, will smart machines, bots and systems powered by artificial intelligence be designed to allow humans to easily be in control of most tech-aided decision-making that is relevant to their lives?

Fifty-six percent of the experts said that, by 2035, smart machines, bots, and systems will *not* be designed to allow humans to easily be in control of most tech-aided decision-making. Forty-four percent said that by 2035 smart machines, bots, and systems *will* be designed to allow humans to easily be in control of most tech-aided decision-making.

Basically, this study proved that we don't know.

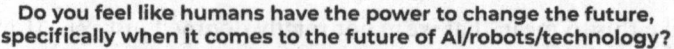

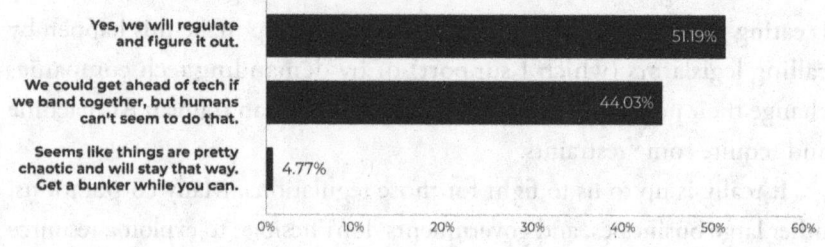

In that Pew Research Center paper, Alf Rehn, professor of innovation, design, and management at the University of Southern Denmark, said it more succinctly than I just tried to: "Non-transparent AI can whittle away at human agency, doing so without us even knowing it is happening."

Those on the other end agreed that humans and tech naturally evolve and usually benefit people most of the time. They felt like regulations and refined ethics would emerge. This means they're assuming we'll get it together, call for regulation, and enact important protections. I really hope this is true, but it's hard for us to imagine something we've never seen before.

What I know for sure is that humans will need to forge a new relationship with technology. We'll need to use it to enhance our human experience, and we'll need to create systems that help us take back our power. Based on our current relationship with technology, I'd say we might have to do this against the wishes of those wanting to keep us without agency.

CHAPTER 8

We're on Our Own

I don't like to talk about my childhood. I hardly remember it, really. I think I blocked a lot of it out. My parents were both alcoholics and later got into drugs, so you can imagine that taking care of me wasn't their first priority. You would be shocked if you knew the things I saw by my seventh birthday. Sometimes my house was filled with strangers. And let's just say they weren't always wearing clothes. I think I blocked a lot of it out because I went somewhere else during those times. I would sit in the closet and create my own worlds. I had my dinosaurs, and I closed the door and did my own thing. It wasn't a closet, but it was the sea or the zoo or whatever. Nobody cared what I was doing. They were happy I was out of sight. I ended up living with my grandparents after that. I remembered being sad I would be leaving my closet behind, but I remember my grandma assuring me that I could have a closet at her house. I did get a closet at her house, but very slowly I learned it was safe to come out into the open. My mom ended up dying of an overdose when I was twelve. And my dad got clean, I think. I don't see him much. I am still in shock that someone could be okay with having a kid and then just not caring at all what happened to him.

—Anonymous, audio-recorded story

Have you ever had a Wine Chip? According to their website, they are "paired perfection" because they pair with wine better than anything else. They're chips that taste like cheese. You can bring them along on a picnic, no refrigeration required.

Wine Chips are the invention of my friend Jonathan, who had a crazy idea to patent a new kind of chip that pairs better with wines than regular chips and even replaces cheese. He went out and talked to cheese people and flavor people and chip people and figured out a way to create a cheesy chip that would complement a glass of Pinot. He rented out a kitchen and began manufacturing these patented chips, nine ounces of which sell for $26. That's a lot of money for a small bag of chips.

But my friend is smart. My friend is successful. His Wine Chips are selling more like hot cakes than Wine Chips. Why?

Meta—Facebook and Instagram.

That's it.

Back when he started his company, Jonathan was spending about $300 a day on Meta ads and then selling about $3,000 worth of Wine Chips.

If you haven't seen an ad for Wine Chips, you probably aren't listed in Meta's database as someone of age who has a higher than likely probability of partaking in a glass of Pinot.

Why am I talking about Wine Chips? I'm hungry.

It's also my friend's company, and one that I know relies nearly exclusively on social media to market their product. There are tens of thousands of other companies who are doing the same.

Ten billion eyeballs scrolling social media for an average of two to five hours a day makes a pretty large pool of potential customers who might be willing to splurge on an extra cheesy chip that pairs well with Cabernet. If my friend is spending $300 a day (or an average of 10 percent of his revenue, which is considered a normal marketing expense), think of the millions of entrepreneurs or large companies who are also spending 10 percent of their millions and billions. Meta, with its suite of apps that include Facebook, Instagram, and WhatsApp, reported $140 billion in earnings in 2023 and $160 billion in 2024. We throw around that billion-dollar number lately, but it's an obscene amount of money, and mostly from selling ads.

Marketers, brands, and service businesses are used to spending in order to gain customers. They find it is worth it to spend a little to gain a new customer. This is called the customer acquisition cost. It's the price tag for making someone go from thinking about your product to actually paying for it. It includes all the ads, the person to write the email, the photo shoots, marketing expenses, and . . . data. The more data a brand has, the easier it is to reach specific customers and get them to buy. E-commerce businesses spend between $50 to $500,000 to land each customer they convince to make a purchase. This tends to pay off depending on the cost of your product and whether that new customer will come back for more.

This explains why the data brokerage industry is suddenly so successful. Every brand wants to pay for exact data so they don't waste their time and dollars on marketing to the wrong people.

Businesses love that helpful data. Tech giants love it too. It works for them and earns them a lot of money. Large corporations relying on social media–based profits don't want social media or data collection or loneliness or capitalism to change. Thousands of businesses count on Facebook tricking us with special email notifications, sounds, games, and "likes" that keep us coming back to our phones to see more of their targeted ads.

Why would any of these tech companies want to change anything?

Sure, the issues I listed in the book required that I mention teen suicide three times for three separate issues. But why would tech companies want to temper their sales or lay off some of those temptation tactics, essentially helping alleviate the harmful effects they're causing, if it also means they'll lose money?

Despite every bit of research that shows the harm that social media, screen usage, and the loss of agency are doing, it is still difficult to advocate for regulatory changes.

Mark Zuckerberg, executive director of Meta platforms, was presented with data that exposed Meta's dangers. He knows that his algorithm intentionally works to inflame users by choosing content that elicits anger. He knows persuasive design is like catnip, convincing people to come back to a post, page, or ad, even if it's not in their best interest. He knows his products are causing a mental health crisis. He knows children are influenced by Instagram's algorithms, which are sickening them by triggering insecurities, compulsions, addictions, cutting, anorexia, and suicidal ideation.

Frances Haugen is a whistleblower who released thousands of data points and emails that prove the founders of Meta knew how much they were harming children and proceeded anyway. Haugen explains the "parental controls" and "safety" products released by Facebook have no accountability because there is no data measuring the effectiveness of these products. They are simply PR tools to let parents think their kids are safe with Meta.

A lawsuit brought by forty-two state attorneys general alleges Meta was aware that children as young as six years old were using Instagram, that self-harm content runs rampant on the platform, and of the "deleterious effects on teen self-image." The suit states that Meta not only marketed to children but used addictive tactics to keep them coming back. Meta is not alone. Google-owned YouTube paid $170 million to settle government and state claims that it illegally took data from users under thirteen.[1] Even though the company claims users must be thirteen to sign up for Instagram, Meta's own survey data allegedly found that 22 percent of children aged six to nine and 35 percent of children aged ten to twelve had used the platform.[2]

All this is to say that it's really all about the money. Money can do a lot of damage.

Julie Scelfo, a journalist, pissed-off mom, and executive director of Get Media Savvy, a nonprofit initiative working to establish a healthy media environment for kids and families, explained, "Kids are dying and nobody is changing anything."

She said something that felt like a punch to my gut. "Social media apps are not a necessity. They're entertainment. They're supposed to be fun, like a toy. Now imagine there was a toy with a slight chance of making your kid suicidal, bulimic, or just plain depressed. A toy that could expose them to porn or, worse, a pedophile. You'd never buy it. Yet Zuckerberg's company has created exactly that product."[3]

Nobody with the ability to make big changes wants to work to change it. How much are we willing to accept and how much are we willing to fight?

According to Mustafah Suleyman, the technologist at Google who developed the AI DQN that learned how to learn through the Atari game *Breakout*, there has never been a new wave of technology that humans have been able to decline. Humans may have regarded advances such as fire, the engine, or cars as dangerous or potentially harmful to humanity. They may have been reluctant to try them at first.

Yet, we've rarely been able to say *no* to a technology and keep it away from our communities. We just go ahead and let the new technologies seep into our lives and change everything, from jobs to pastimes to society. While I would love to ask you to just get off the computer and head to a bunker, I know that's not realistic. Technological advancement is here to stay. And while we can push for legislation that requires corporations to donate or give back or fund programs that help addiction, the bottom line will always be most important to corporations. For them, it's earnings over all else.

The list of regulations that could curb the actions of Big Tech is long, but the lobbying power of those tech giants is great. It heartens me to know that several congresspeople and senators have already introduced bills to help enact privacy laws, laws regarding data minimization, laws that would mandate the labeling of AI-generated content, and laws to protect us from addictive algorithms.

However, not one of the bills introduced has made it into law. Tech lobbyists padding the pockets of government officials and politicians have thus far staved off forced regulations.

Most Big Tech leaders attended President Trump's second inauguration, so I don't hold out much hope for changes to be made at a federal level. Big Tech's net worth makes legislation difficult. Despite clear evidence that something is harmful to American citizens, there are groups with billions of dollars and significant lobbying power making bills to curb harm extremely difficult to pass.

This is why we still have AR-15s despite there being 488 mass shootings in 2024 alone.[4] There is myriad proof of the scope of the problem, yet corporations pay an obscene amount of money to keep things just the way they are. According to the lobbying disclosure data on opensecrets.com, the National Rifle Association spent $1.5 million in 2024 to convince lawmakers not to pass gun laws. And Meta spent $18 million in that same year. That's a lot of money going into a lot of pockets to make sure that Meta and all its predatory practices remain intact.[5]

The White House agrees that regulations are necessary, or at least they agreed before 2025. President Biden's Office of Science and Technology Policy published a *Blueprint for an AI Bill of Rights* in 2022. It set out to be a "set of five principles and associated practices to help guide the design, use, and deployment of automated systems to protect the rights of the American public in the age of artificial intelligence." It was an interesting attempt to set up some sort of framework for businesses as they trained their machines and set up new AI systems. Someone in that administration must have said, "Hey, this will never go anywhere with regulations so let's at least do something."

Few people heard about this blueprint. I am a tech founder and didn't hear about it until two years after it was written. It was deleted from WhiteHouse.gov on the very day President Trump took office. It's nice to know there was at least an attempt to protect the American people.

The awkward TikTok ban in early 2025 lasted just a day. Those suggesting the ban were accusing TikTok of posing a threat to national security, and so the platform was shut down the day President Biden left office. The next day, it was up and running again with a confusing message about how its reinstatement was all thanks to President Trump. The controversy over TikTok as a security threat seems to have since disappeared. However, in the wake of the ban, Congresswoman Alexandria Ocasio-Cortez made a video that slammed the entire process. She said that there was no compelling evidence that proved TikTok posed any threat to national security. She went on to explain that the lobbying power of Big Tech is causing lawmakers to play endless "whack-a-mole with apps" instead of focusing on the bigger issue, which is privacy laws. She argued, adamantly, for more privacy legislation and for individuals to have greater agency over their personal data. In short, Ocasio-Cortez explains that people deserve not to have foreign or domestic companies spying on them all the time.

I agree with her statement: "The degree to which corporations surveil the American people without genuine consent and without real mechanisms for opting out should be illegal. Point blank. Period. The answer is that we should have real privacy legislation in the United States, and we should let people own their information and have greater agency over their personal information."

Agency over their personal information? That sounds like a statement I would make! And though Congresswoman Ocasio-Cortez agrees with me along with the several other senators and congresspeople who have put forward several bills that aim to label AI-generated content, set stronger data protection rules, restrict data collection, increase research into the effects of digital media and gaming on minors, and audit algorithms for harmful biases, every single bill has been met with opposition from Big Tech (or the lawmakers who receive big donations from Big Tech), and nothing has passed.

The sad truth is that we cannot trust that large corporations have our best interests in mind or care about the harm they are causing. We have to fend for ourselves. Or . . . we can devise a plan that helps business just as much as it helps humanity. And I think I may have done it.

The story economy helps us enhance our human experience through technology. It ethically collects data from our stories and allows us to use it according to our needs. It benefits both humans and corporations, so I can imagine wide acceptance of its use. It just might work. But in order for that to happen, we all have to be willing to embrace the discomfort of sharing our stories.

The first step is to learn how much power one story can hold.

PART 2

The Power of a Single Story

CHAPTER 9

Letting It All Out

My dad died and I didn't talk about it for twelve years. I was sixteen when it happened, and my family's way of dealing with it was to sweep it under the rug. We just didn't talk about him. No pictures up. No celebrating his birthday. It was just . . . Dad who? I accidentally said his name once like three years after he died and my mom immediately jumped on me and said something like: Why are you *still* thinking about him?

I went to college and tried to join a sorority but there was this daddy-daughter dance thing we were supposed to attend. I didn't think it was appropriate to just tell them that my dad died because I had no idea the topic was allowed. So, I quit and that was probably a good decision. Thanks for looking out, Dad!

Over the next few years, I worked on holding in my story, but boy that was a lonely experience. I think underneath it all, I wanted to talk about my dad. He'd been my dad for sixteen years. And then he just wasn't, and it was so weird to just forget him. Anyway, so finally when I was twenty-eight I made a friend who was one of those true friends who really wanted to actually know me. And she asked me questions about everything, including my dad.

I felt like I was doing something wrong telling the whole story, but I cannot emphasize the relief I felt once I told it. That story had been fighting so hard to come out, and I truly became a new person once I actually told it to someone else. I began writing about it, and I published one of my stories about his death in a national publication. It went viral. So many people related to it. The simple act of telling my story was helping so many people.

I was like . . . shoot, I should have done this forever ago.

—Anonymous, audio-recorded story

Have you seen the trauma salad trend on TikTok? (Surely a long-dead phenomenon by the time this book is in print.) My daughter, Alaina, showed it to me. It's both captivating and depressing. Friends take turns sharing true stories of things that they went through that caused them trauma. Each time they share, they add a different candy to a bowl to create a mixture of Nerds, Gummy Worms, and Haribo treats—the best kind of salad.

Whenever I see candy now, I think of these stories. I am still mortified and shocked at some of the things that happened to each of the storytellers.

"Hi, my name's Mercedes," one says. "And when I was fourteen, I found cameras that my mom's boyfriend had put inside my bedroom."

She goes on.

Many other teenagers follow with their stories, each contributing a gummy snack while unloading their trauma into the bowl. Some stories have to do with their parents' actions and behaviors. Some speak about their dads going to jail, catching their moms cheating, getting beaten up by a parent. Each one is sort of one-upping the previous teller's stories with a more traumatic childhood experience.

The stories feel authentic, like real truths. The tellers feel vulnerable. I'm happy that they're sharing their stories. I hope sharing provides catharsis, a new perspective, or that comments they receive will help them see that they're not as alone as they might have previously believed. Most commenters send them hugs or ask if they're okay. It's a lovefest under every #traumasalad post.

I love this trend because it means we're heading into more vulnerable waters. People are overcoming stigma and fear and getting more and more comfortable with sharing their true thoughts and experiences.

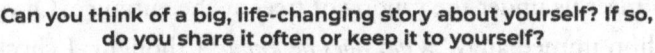

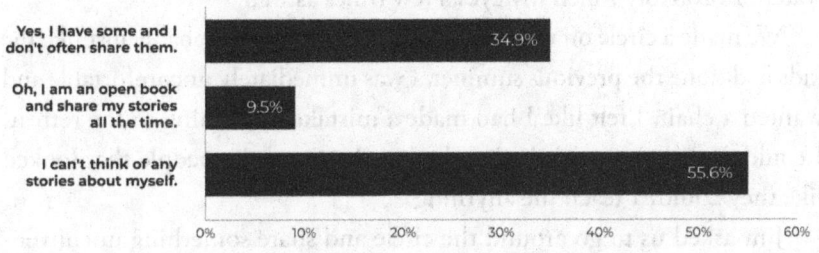

I learned firsthand how powerful vulnerability could be when I was in my late twenties. I knew I wanted to be a better leader. I had left the Air Force for a job on Wall Street, but it happened to be right when the financial crisis hit, and I lost my job right before I moved my whole young family to New York. I was lost. I didn't know what to do next, but I knew I wanted to continue being a leader in some capacity. So, I signed up for a leadership course in the forests of the Catskills.

I drove hours to get there, a little nervous, expecting the whole ropes course thing and some cheesy get-to-know-you activities. I figured I'd make some connections, get some business cards, learn a thing or two, and skedaddle back home. I had small kids then, and I felt pangs of guilt for leaving my wife alone with them. This would be a quick trip and look great on my résumé.

I got there early and joined another guy a few decades older than me who had also arrived early. I judged him immediately as someone who was late to the game. *You should learn this stuff in your thirties, not in your fifties*, I was thinking. I didn't give him the time of day. We just sat there in silence and waited. Other participants began to trickle in. Some were my age. Some were older. There were a few participants in their twenties. I judged them as well. *They must be cocky to think they'll be leading so soon.*

The retreat was in a forest location on the campus of a kids' summer camp. The campers had left to return to school. The leaves on the trees were just starting to turn fiery red, yellow, and orange as autumn set in.

Our retreat leader, Jim, a balding guy in jeans and a cuffed dress shirt, came to meet us under the canopy of trees at the entrance. Of course, I judged him immediately. *What does he know?* I thought. I checked my watch. I probably rolled my eyes a few times as well.

We made a circle on the floor inside the mess hall, probably just like the kids had done the previous summer. I was immediately uncomfortable and wanted a chair. I felt like I had made a mistake by coming to the retreat. I couldn't shake the weird vibe, the location, and the people that looked like they couldn't teach me anything.

Jim asked us to go around the circle and share something unconventional. We weren't allowed to say our names or what we did or our goals or the reason we were there. We had to share something otherwise not known about us. He didn't say the word "vulnerable" but that's what he meant. We all stared into the circle, careful not to make eye contact. Suddenly, the older guy spoke.

"My wife left me yesterday," he said. He said it just like that—no sugarcoating.

"We've been married thirty-two years. We had breakfast just like normal. We talked about our kids. We went for a walk. And then she just said she wanted to live out the rest of her life on her own. Just like that. So matter-of-fact. What hurts the most is not the fact that she's leaving but that she must have been thinking about it for so long. I think about all the times we must have been out to dinner or at the movies or in bed reading or whatever and she was probably thinking, *I'm going to leave him soon.* Yeah, that really hurts."

This boom of silence followed.

Then, a woman spoke. She said she was scared for her son. She didn't know where he was, and she was nervous to be in the forest without service in case he needed her. Her son was a heroin addict, and she was always fearful that every call she got would be the police telling her he was in trouble or dead. She'd sent him to rehab. She'd sent him to live in other places. He always found drugs. She worried she was a horrible parent to him so she could never be a good leader for others.

We continued around the circle, each person going deeper and deeper than the previous sharer. I am not sure that's what Jim wanted, but I think the vulnerability of the first man's share sparked deeper storytelling.

When it was my turn, I wasn't sure what to say. There were so many things that weighed on me. What came out was surprising. I began to speak and admitted for the first time out loud that I was scared I was disappointing my father and my family by not continuing in the Air Force. I told all these strangers how I was scared to tell my family what I was really doing and what I really loved. It was just then—at that very moment—that I realized I was scared of disappointing my father. I was a captain. I could have had a great life just like my siblings and my father and so many others in my family. I chose to take a risk on a new opportunity, and it had failed. I had failed. I was flailing, and I knew everyone in my family was going to shake their fingers at me and tell me I had made a stupid choice.

The entire weekend revolved around these shares. Every single one of us cried. We weren't learning tricks of the trade. We weren't learning how to deny a raise or hold a good meeting. We were there to learn how to see people as more than their appearances. We were there to see how so many of us hold such important stories inside. We were there to learn more about people and how to treat them, knowing the gentleness and importance and vulnerabilities they're holding just beneath the surface. This was a different, more humbling kind of leadership. I didn't expect it, but this weekend changed me fundamentally, more than I could have imagined when I had just arrived there with all my judgments.

I left the retreat with more than business cards. I left with a heart filled with a lot more compassion than I had before, and a new way to connect with people. While we've not met as a group since then, we are forever tied together by that week, and by those stories. Those stories connected us. Had we only shared our job titles or our salaries, we would have never felt so close. Yet, it's not in our culture to start out that way. I think the first man, who shared that his wife had left him, did so because he was too broken to hold up the barriers we humans usually use to divide us. Imagine what would happen to the loneliness epidemic if it *was* common to simply share the truth at first sight.

The act of sharing can be freeing to you and helpful to anyone who hears your story. Laurenne Sala, along with her cohost, Corey Podell, produced the monthly Los Angeles–based storytelling stage show *Taboo Tales* for ten years during which she helped hundreds of people craft essays and then share them onstage in front of sold-out audiences.

Often these were stories that may seem taboo in society. A list of topics covered on her stage includes: sexual assault, plastic surgery, incest, stalking, assault, living in cults, religion, cancer, HIV, sexual experimentation, and abortion. These are stories that people hold inside for fear of shame, judgment, or retaliation.

I asked her what she's learned over ten years of telling and hearing taboo stories, and she said, "People desperately need to share who they really are, but the world tells them not to. They say in the recovery world that 'you're only as sick as your secrets,' and I believe that to be true. I have watched countless people take to the stage and let go of all their shame as their story comes out. You can almost see it leaving their bodies. It's so freeing."

I was able to catch a *Taboo Tales* show myself years ago, and I remember feeling a sense of calm as I heard others share stories that were just as taboo as the ones I knew I was holding inside.

Speaking a story aloud is powerful, and everyone who hears it feels connected, Sala explained. I'm fascinated with her motto: The more we talk about how fucked up we are, the more normal we all feel. It's interesting to me how rare it is to feel normal. Does anyone? I love that hearing a story can make you feel normal, or like someone somewhere understands what you've been through.

There are several places to hear true stories now. *The Moth*, *We're All Insane*, *Terrible, Thanks for Asking*, and *This Is Actually Happening* are podcasts where storytellers share their secrets and truth. Each can provide you with a sense of belonging, or normalcy.

I've listened to them all, and they always help. Each time, I begin thinking that I won't relate to the storyteller, and each time I find I have something in common with them.

I listened to a podcast recently about a gang member who went to prison for murder. He got out after years of good behavior and started a successful cleaning business. As an Air Force brat and engineering student, I was never close to joining a gang or murdering someone. Yet, I felt so close to this guy who shared his sentiments about wanting to learn from his past and make something good out of his life. There's such a magic connection in the telling of a story.

However, in order for all that to work, stories have to be honest. The magic that comes from sharing a story doesn't work unless the story is real. I may tell a story about having a difficult time with my teenage son. My story may inspire you to share a story about your teenage son. If you later found out my story was untrue, you'd feel betrayed.

Lying to oneself doesn't usually help matters either. If I lied to my therapist, like I've heard many people report doing, I am fairly certain I wouldn't have as many epiphanies.

We must be honest in order for the telling of a story to do the hard work of revealing who we are.

How do we know if we are being honest?

What kind of stories are we telling? Our culture isn't really so conducive to being honest with ourselves or others. We answer "fine" blindly when people ask us how we are, and we often try to believe it ourselves.

There may certainly be cultures or communities where people are more supported in telling the truth, but as a rule, human beings are prone as a species to being big fat liars. It's been this way since the beginning.

Humans are the only species to create multiple personas—one for personal use and others for various public interactions.

My Aunt Sally used to cover her living room sofa and lounge chairs in plastic sheeting. This was the main room where she entertained people. She did this to demonstrate to guests how clean and perfect she was. In reality, her house was a mess once you left that pristine room. Don't even ask about the towels and soaps we weren't supposed to touch in the guest bathroom. There were always two stories at play in that house: a real one and an aspirational one for public consumption.

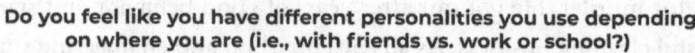
Do you feel like you have different personalities you use depending on where you are (i.e., with friends vs. work or school?)

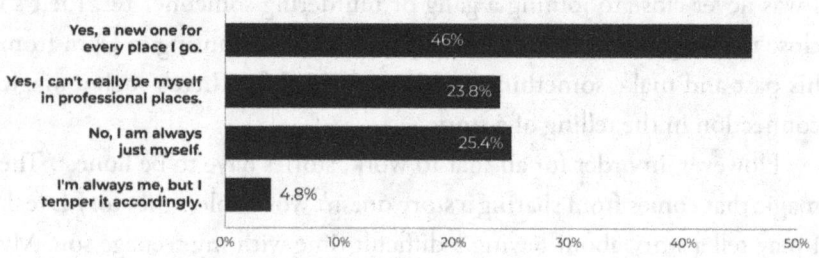

- Yes, a new one for every place I go. — 46%
- Yes, I can't really be myself in professional places. — 23.8%
- No, I am always just myself. — 25.4%
- I'm always me, but I temper it accordingly. — 4.8%

When I think about the public persona, my mind immediately imagines the green lawns and smiling housewives of the 1950s. While I wasn't around then, every idyllic movie about that time shows a happy family, perfection in the kitchen, separate beds or bedrooms for husbands and wives, and a dad who easily brings home the bacon. All these scenes hint at secrets and flawed lives off-screen. Nobody seemed to tell anyone the truth back then. They all appeared to be living in some version of *The Stepford Wives*.

I asked my parents about this time period. My mom reminded me that there weren't many divorces back then but there sure were lots of unhappy women. They just never said it out loud.

Half of marriages end in divorce these days, so maybe honesty is spiking. I guess that's good, but we've still got lots to work through.

Caring about what others think about us and creating a public persona that pleases those around us goes back thousands of years. It's one of our most useful and common defense mechanisms. Back when we lived in tribes who looked out for each other, it was extremely important to stay within the tribe. If a person was shunned or ostracized from the village, the consequences could be catastrophic. You might not be able to hunt or gather food. You wouldn't have a warm place to sleep. You wouldn't have the conveniences of sharing in all the necessities that a tribe provided. Being exiled could mean hardship, loneliness, and no protection from other tribes or animals. Being shunned could mean death.

It didn't happen often. However, there were certain times where ostracism was necessary. That threat might have made some tribe members keep their vulnerability to themselves.

Certain tightly knit groups who remain still use ostracism as a tool. The Old Order Amish, for example, use shunning as the most severe punishment for the most serious offenses, such as marrying a non-Amish person, adultery, excessive contact with the outside world, and drunkenness. Shunning is instituted only through a unanimous vote by the church community.[1]

While times have mostly changed, the feeling persists: that we needed to be included and liked in order to survive. To many people, sharing a vulnerable story still feels like a life-or-death situation. Share who you really are, and your entire community might tell you to get out.

While actual shunning is rare, "being canceled" is common, and it's not that different. Say the wrong thing, and the entire internet is emailing your boss and asking for your resignation. I blame a lot of this on the arrival of the twenty-four-hour news cycle. Social media hadn't yet hit the scenes, and reporters were desperate for stories. When cable news came on the scene, networks needed to create coverage over a much longer span of time, so they found anything to fill the slots. Someone fell off their bike? Let's tell everyone at 5:00 p.m. Someone is mad at a neighbor for picking some fruit off her tree? Let's spread the word and start a series about neighbors stealing fruits. It was great for the news networks, but it has made dropping our public persona almost impossible. Nobody wants to be humiliated in front of a big crowd.

Being ostracized now has global repercussions. You can have your dog off its leash in Manhattan in one moment and someone in Korea can be watching you on a TikTok video and be disgusted by you the next. We've made it scary to do anything. We might get sued. We might end up on the news. Everything is being recorded. Sure, the world might be safer with more caught on camera, but it can feel torturous to simply be yourself. I've seen sneaky content creators recording innocent people with imperfect bodies at the gym only to get laughs. Everyone is in danger of ostracism these days. This cycle plays to our deep desire to be liked that has been surfacing in us since we were tribal. It's in our DNA.

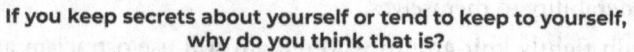

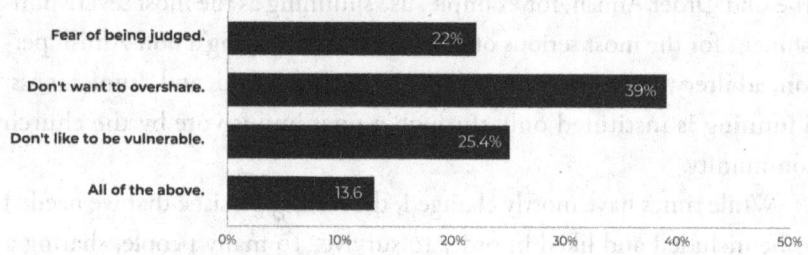

We've retaliated by starting our own tribes. Creators often find like-minded people and begin gathering followers. Their leadership role often inspires instant likes and social validation. It may feel great, like a relief. Online communities grow then, and the stakes get even higher. There are more people to please and brand deals to secure. There's a pressure to be liked and to keep your mask on for just a little longer.

How do we navigate all this? We lie. We perform.

In 2019, the *Harvard Business Review* reported on the phenomenon known as code-switching, which they define as "adjusting one's style of speech, appearance, behavior, and expression in ways that will optimize the comfort of others in exchange for fair treatment, quality service, and employment opportunities." HBR explains that code-switching has "long been a strategy for black people to successfully navigate interracial interactions and has large implications for their well-being, economic advancement, and even physical survival."[2]

They illustrate clear evidence of code-switching with a viral video from 2012 of Barack Obama congratulating the US Olympic basketball team after a win. He gives the white coach a polite, formal handshake. Then, when it's time to shake the hand of Kevin Durant, a Black player, he gives a very different kind of handshake that seems more casual. The video inspired a code-switching sketch by Key & Peele that pokes fun at the whole idea of the switch.

For many Black employees in corporate America, code-switching feels necessary to survive and to climb the corporate ladder. The *Harvard Business Review* interviewed three hundred Black people employed in American companies, and the average rating they gave themselves was a level 4 of 7 in code-switching behavior, meaning that code-switching is a reality for everyone in the survey.

A woman in her thirties who works as a senior research program coordinator told the *Harvard Business Review*, "In my actions and verbal communications, I try to avoid any opportunity for someone to label me as the 'angry black woman.' I also carry myself in a professional manner that may seem to be a step above the somewhat casual professional environment of the office."

I admit that I can't know what it's like to feel the pressure to be a minority trying to survive in an often white-led corporate world, but I felt called to explore the idea of code-switching because I often feel the need to do something similar. As the founder of a tech company who is responsible for the culture of the place, I feel as though I must behave in a certain, almost neutral way to please everyone simultaneously. I meet with the entire company every morning at 9:00 a.m. and that means sixty different personalities are looking to me to start the day. There are eccentric creatives, brainy engineers, funny behavioral scientists, and serious UX designers. Not all of them appreciate my humor, and I'm getting used to silence on Zoom calls. When I meet investors to ask for money, I have to wear a different kind of professional mask, hoping they see me as dependable and brilliant. When I get home to my wife and kids, I sometimes don't know how to act anymore! It's exhausting.

Of course this carries over to how we act online. Many of us can't help but put on a persona there as well. Everyone knows Instagram was created specifically to show off how great your vacations are, the amazing food you make, and your spotless pantry. Many people see themselves as a personal brand. This comes with pressure to adhere to those brand guidelines. There's not much room for error—only perfection and judgment.

According to the UK's Custard.com study, only 18 percent of men and 19 percent of women reported that their Facebook page displayed "a completely accurate reflection" of who they are. Most commonly, participants said that they only shared "non-boring" aspects of their lives (32 percent) and were not as "active" as their social media accounts appeared (14 percent).[3]

Most of us have seen articles about someone getting caught for lying on social media. There are filters, luxurious backgrounds that fake vacations, celebrities constantly editing their photos so they look perfect, showing off stacks of fake money, clean houses, perfect marriages, muscular bodies, and organized pantries that have been debunked in other posts or simply cannot be real.

We are not telling true stories even though those are the stories we need to tell. Even the trauma salad trend isn't totally honest. When I watched TikTok creator @Olive talk about the horrible things that had happened to her during her trauma salad share and then laugh hysterically, I was confused. Was her laughter cathartic, because although the situation was profoundly painful, it was so dark it was downright demented? Was she telling the whole truth about how she was feeling? How does she really feel about it? We might never know.

I get it. We're entangled in a society where truths are evaded, and people are encouraged to do things like "Fake it until you make it." Validation is often hinged exclusively on money or status. Status in this case is not necessarily earned through real achievements or pro-social behaviors like kindness or compassion. We praise people for their physical beauty or black credit cards or Lamborghinis rather than for what they truly contribute to the world. The people we praise for their accomplishments are often charismatic, silver-tongued fast talkers—all pomp and no substance. Sometimes we are misguided into thinking we aren't good enough, that if we pretend to be someone we aren't—someone prettier, stupider, more charismatic, more together—for long enough, we will be accepted.

This society can also batter our true selves. Feigning indifference, emotionally fleeing situations that trigger old traumas, twisting our days and lives up in knots in order to be financially solvent can wreak havoc on our psyches. If the corporations control the false narrative and we need to make money to continue to eat, it makes sense we might invent influencer types of personalities to attract the right advertisers. If our false personas were developed as a means of survival—as a way of existing in the world without making too many ways and freaking others out—it can be difficult to know how to be any other way.

Being our true selves can help us and set us free.

Brené Brown spent two decades studying courage, vulnerability, shame, and empathy. Brown believes that you "have to walk through vulnerability to get to courage."

We have to, at some point in our lives, tell the truth. It might not be when we're trying to climb the corporate ladder or needing to earn money as an influencer who must pretend to love Amazon clothing. However, someday, if we want to learn more about who we really are and if we want to truly connect with other people, we're going to have to choose honesty. We have to tell our stories.

It will be uncomfortable, but we must be honest about ourselves and face our stories head-on to find the strength in them or rewrite their endings. This is to say, we have to take off our masks. Even if it feels scary. Even if it feels like removing our masks will take away opportunities. It's the only way forward. Because hiding who we truly are is detrimental. And life is too short to be lying all the time.

CHAPTER 10

The Stories Inside

I am so embarrassed that I was catfished. It's honestly something I would never tell anyone besides an anonymous voice recorder on my phone. Like, please do not ever figure out who I am. I work in social media marketing. I am quite savvy online. I don't know what the hell happened to me. It was such a lapse in judgment. I guess he or she, or whoever the hell it was, caught me at a very lonely, vulnerable moment. I won't tell you everything, but they sent me some steamy videos, and I sent some back and we sent each other voice notes all the time all day long for months and months and I'm over here thinking he was going to just finish the job he was on and then we were going to finally meet in person and maybe get married and who knows? I was so excited. Anyway, so this supposed job ended but then he had an emergency with his mom he had to attend to and then after so many failed meetups, I realized this person was never ever going to meet me ever. And it was so hard to break it off despite knowing that. I had really connected to this person emotionally. We had seen each other's bodies. And now I have no idea whose body I actually saw and who actually saw mine. It's just so embarrassing. And, honestly, I'm still heartbroken about it.

—Anonymous, audio-recorded story

Just over two years ago, I was meeting with a group of colleagues who were brainstorming the next steps in building out my company. We had traveled from across the country to be together because, well, apparently what happens when you start a company during a pandemic, people can basically live anywhere. The talk around the table revolved around how much air travel had changed and everyone was filling their coffees and clanging the table with caffeine-filled mugs. We launched into discussion after discussion. Then, somewhere around 3:00 p.m., everything went blank.

Totally blank!

I lost everything. All my thoughts. All my words. I saw nothing. I heard nothing.

I was gone.

Everything was black for what seemed like an eternity to me.

Then, just like a computer being rebooted, everything was back again. Yet, nothing was ever the same.

Nobody in the room knew that I had just experienced a blackout. My body hadn't dropped to the ground. I have no idea what my face did. The whole ordeal must have lasted a few seconds, but to me it was forever. I searched everyone's faces to see if they had noticed. I spent the rest of the meeting trying to remain calm when I knew everything was different. I was able to make it through the session and even offer a technological solution to a major issue we'd been grappling with. Still, as everyone walked away from that meeting, my senses were numb.

When everyone left the conference room for the day, I stayed for a long time trying to grapple with what had happened. It was like my software updated, and I became a totally different person.

People have told me I had a spiritual awakening. I don't know what to call it. I basically came away with a greater knowledge that many of the things I was doing in my life weren't healthy for me, and I knew I needed to make big changes. Everything was fine one moment and the next, well, I was estranged. I didn't recognize myself.

I made it to my trusty Toyota 4Runner and started shaking and crying inconsolably. My plate-spinning, flame-juggling, orchestra-conducting self had watched everything fall and shatter in an instant. If I were a machine, I would have become William 2.0.

Maybe this sounds normal to you. Maybe you're already on version 3.3 of yourself. I was stunned. I'm still dealing with the fallout. It's like I woke up from that three-second blackout and I was suddenly aware of every change I needed to make in my life. It felt like I was given this moment to really take a step back and do something. It felt urgent yet confusing. It might even be the reason I'm writing this book right now.

I stopped drinking alcohol that day. I let go of certain relationships. I began taking better care of my health, and I started going to therapy. I knew from that day forward that I had to help myself in certain, bigger ways.

I had seen the research from Brené Brown. I understood the power of sharing vulnerable stories aloud, but I hadn't truly done it myself in a big way. I had tried therapy here and there, but this time I knew I needed to really take it seriously and tell my true stories. I wanted to figure out what the hell happened to me that day, and to understand how I could use that information going forward. My inner self was in shambles. I needed help.

TALKING IT OUT

My new therapist Rebecca's office felt like such a stereotype: calming music in the waiting room, boxes of tissues, a few very outdated magazines. When Rebecca came to greet me, I felt a sudden sense of calm. I knew this was the right thing to do.

When we spoke, I learned she was open to seeing me for who I am and agreed not to feed me a therapeutic formula and guide me through a workbook. She would listen to me at every session, and we'd go from there. In turn, I would be absolutely, 100 percent honest.

It worked, and it's still working.

These therapy sessions where I consistently talk it out have truly transformed me. I am more aware of things that have been haunting me my whole life. I deepened my understanding of the fear of disappointing my father that I copped to in that circle at the retreat. I learned why one of my defense mechanisms is to judge other people to attempt to evade intimacy. I learned about my deeper truths and wants.

I feel like all the awareness that came to me that day in the office was just the beginning of a new me.

I truly am William 2.0. The fourth. Or would it be William 4^2? I'm not sure.

The most important thing Rebecca helps me with is to identify the *inner stories* I've been holding inside. Inner stories are ideas we accidentally believe that sometimes drive our decisions. We have inner stories about ourselves, about the world (unconscious bias), and about our beliefs (cognitive bias). At some point, any of them can run the show, make decisions for us, and close us off from opportunities we didn't even notice because those stories got in the way.

I had a cultural bias common to many men in my generation: an inner story we first learned when we were children. It goes something like this: Men are failures if they don't provide for their families.

Does this one sound familiar to you or anyone else you know?

That story has affected a lot of my decisions over the years. I had to hold off starting the story economy until I was sure it was a viable business plan, and that the business had real potential to provide enough money to support my family, especially with my three kids who would eventually be leaving for college.

My inner story would have pushed me toward feeling that pressure even if my wife, Eileen, had told me not to worry about bringing in an income for a few years. That story is so ingrained in me that I probably would have made a similar decision to wait, save, and be sure of funding, even if she were the CEO of Apple.

While it might seem like I had made a good financial decision, the story that drove it has caused me a lot of harm.

I put an exorbitant amount of pressure on myself to scrimp, save, and make sure everyone is okay. Rebecca has guided me to understanding that storyline has prevented me from asking for help. I try to prove to everyone that I don't need them, that *I* am the provider. That can be tough.

This storyline was also causing me to devalue my wife. She might come to me proud of one of her accomplishments or excited to talk about a trip she had planned and paid for, and I would feel like a failure for not having paid for it myself. This little story has been part of the fabric of my belief system for so long that it's taken me some forty years and hours upon hours of talking it out in order to dissect it and learn from it.

Eileen actually shared one of her own story traps with me. She says that many women devalue women's worth after fifty. She believes this is why our culture is so hyper-focused on youth and looking young—because the moment you look old, you're ignored, discounted, and treated like you don't matter. One thing she may be working on in her own therapy sessions is to be okay with the idea of feeling pushed to the side within society as long as she feels important to herself.

Whew, the progress we're making in therapy is monumental. It's all about figuring out what kinds of stories we allowed to shape our personal and shared realities.

We learn many of these false storylines before we are even four years old. They might not have been taught to us in a class. We learned through those nonverbal cues that Marcos Pantoja, the man raised by wolves, didn't get, which communicate so many ridiculous stories we ended up internalizing.

Stories aren't necessarily good or bad, and there aren't necessarily any "right or wrong" stories. They are simply stories that help us make decisions in the future or act in certain ways as adults.

Some people with critical parents might learn in childhood that they "can't do anything right." Carrying that story can be harmful as an adult. If you believe that, imagine how you might act at work. You might never speak up in meetings or might present a PowerPoint with apologies, assuming whatever you say won't be good enough.

On the other end of the spectrum, you might have learned a story in childhood that "you do everything right." Someone challenges you at work and you might get defensive or dismiss them without considering their opinion. Either of these stories might hurt your ability to do your work unencumbered, to align with your true talents and purpose as you move forward in your career.

These inner storylines are just as powerful as the ones we tell out loud. It's really important to figure out what kinds of stories we are holding onto deep inside our subconscious in order to move forward in all aspects of our lives.

Rebecca taught me that we can't really understand our *inner stories* until we tell our *outer stories*.

I've been taking off my mask and telling Rebecca my true stories every single week for five years. I have a new routine. I go to therapy. I talk it out. Maybe I cry. I usually have a pretty good realization. I figure out a new inner story that has been running the show. I exhale with relief. I go treat myself to a dirty chai with oat milk at my favorite coffee place in Austin while sitting with their two resident goats. I think about everything I just shared and what it all meant. Then, I go on with my day. It's changing my life.

NEUROPLASTICITY

I've since learned, having these realizations actually changed me physically. Psychologist Thania Siauw says, "Talking therapy changes the brain. We are able to increase blood flow to the cortex by sharing our experience with another person—as this increases our capacity to regulate emotions, attune to others, self-reflect and better express ourselves, all of which are associated with mental wellness."[1]

Basically, she's saying that the very act of telling our stories can help us. She goes on to say that "talking can support your brain to make sense of painful memories in a different context, and can help neurons from different parts of the brain communicate more effectively. This is a significant part of the healing process, especially if you have been through trauma."

I have seen this firsthand. Rebecca is masterful at guiding the conversation toward the epiphanies I experience. I also have similar breakthroughs when I'm simply sharing my story into a voice note. Talking out loud can help me see what actually happened from a new point of view. And playing it back later has an even more profound effect.

SEA SNAILS

Eric Kandel won the Nobel Prize in 2000 for his studies of the Aplysia, a type of sea snail with a simple nervous system. He was trying to understand how memories are stored in the brain. He discovered that as the Aplysia learned, chemical signals changed the structure of the connections between their cells, the synapses. These synapses are what the brain uses to communicate. Basically, Kandel proved that learning new things, even about yourself, can truly change how your brain works and communicates. It's not that your memories are stored somewhere and come out to remind you that the new gas station is coming up on the left. It's that your brain physically changes with this new information.[2]

The simple act of learning or revealing the story that you picked up in childhood about how you "can't do anything right" changes your brain. Sometimes the alteration is drastic, in that after releasing that storyline your mind no longer believes it. Often the changes are more subtle. Your brain may communicate with you slightly differently the next time the storyline surfaces in your life.

We might talk out loud and realize that we have a pattern of getting angry in traffic. It might take us several times telling that story to realize something is happening we don't like—we can't stand when people are rude in traffic, and we end each drive in a sweaty rage.

After having that realization, we can ask ourselves why we are repeating the storyline. When we become truly curious about ourselves, our past, and our stories—we are closer to finding a real answer.

We may get cranky in traffic because our parents placed respect high on their value list, and being cut off in traffic is the ultimate form of disrespect. Or perhaps our parents may have taught us that being late is not acceptable, so we get anxious in traffic. You may have an odd trigger. Maybe you once watched that episode of *Knight Rider* where KARR, Kitt's alter ego, ran amok and it terrified you, so that every time you are trapped on the highway you worry about lunatic self-driving cars.

The moment we discover the stories that we have told ourselves repeatedly is always a trip. That story that had lodged itself in your mind for decades, really caused you some issues, and suddenly you are aware of it and can put it away for good.

When we share the story, think about it, process it, and reflect on our experiences, we can learn more about ourselves. We gain the ability to figure out all the stories we've come to believe and ask ourselves if we still want to believe them or what we want to do with them when they reveal themselves in our day-to-day lives.

This new information allows us to change the neural pathways in our brains and teach ourselves to have different reactions. We can teach ourselves to not get so angry in traffic or to not accept low offers for our valuable time. We can teach ourselves to stay calm when our kids are having tantrums. Whatever the habit is, we can talk about it, figure out its underlying story, and work to change it.

A study published in the *European Journal of Social Psychology* is often cited for demonstrating that new habits take anywhere from 18 to 254 days to form, with an average of about 66 days. In that study, 96 people were asked to choose a new health habit and practice it daily for 84 days.[3]

Of the original 96 participants, 39 (41 percent) successfully formed the habit by the end of the study period, proving it's totally possible to change your brain. For example, if you simply notice your pattern of getting angry in traffic for 84 days and don't act on it, there's a 41 percent chance (or more depending on if you go as far as 254 days) that your brain will actually be changed to be conditioned to no longer act on your road rage. That's pretty powerful.[4]

It all starts with a story.

You might already know this if you have kids, but it can be impossible for someone to learn a lesson without firsthand experience. You can tell your kid a million times not to lean back in her chair at the table, but she might not stop until she finally falls. Similarly, we can read a million times about how road rage is bad, but we might not stop our habit of flipping people off in traffic until we realize that our habit of dealing negatively with our deep feelings of shame centers around being disrespected.

The human brain (like the sea snail brain) is quite a marvel. There's so much going on inside our own minds we aren't aware of. Taking the time to question, observe, and identify how it's working can change everything. It's a worthwhile endeavor.

We have the research that telling true, vulnerable stories out loud can help us create deep connections and even change the neural pathways in our brains. We know storytelling can help us see ourselves in a clearer way. Yet, we tend to shy away from sharing. Do you tell honest stories? What if I challenged you to do it? What if I told you that there are other, sneakier stories lurking inside of us that we all carry—and that sometimes they make bad decisions for you? If you knew that sharing your truth would help you uncover those, would you try it?

CHAPTER 11

Cognitive Bias

I don't know why but I never ever wanted to date Korean women. It must be some deep-rooted shit I learned as a kid. Or maybe I was running from people like my mom. I didn't want to end up with someone like her, totally mean and controlling (sorry, Mom). I have no idea what the reason was, really. But on the apps, I just automatically swiped left on any type of Asian woman, really. Stuck to white and sometimes Black women. That's it.

My brothers gave me shit for it. My friends did too. My cousins. Everyone was always laughing when I brought a white girlfriend around to a family party. And you know what? I thought I had it all figured out. They would laugh at me, and I would just let them, thinking they were the ones missing out.

But . . . you know where this is going . . . one day I sat next to a girl on a plane, and we talked the whole time. She was a beautiful Korean goddess. She looked like some statue you'd find in a museum in Seoul. She held herself with such confidence. I was almost a little scared of her. But after we talked a bit, I felt so comfortable talking to her. We had so much in common. We both grew up in Southern states but moved to Cali for college. So, we both were called the same racist names as kids by the same kinds of people. I always thought it would feel boring to date someone who had a similar kind of upbringing, but I can't explain how right it felt to just kind of know someone right away without knowing them. Like, I wouldn't have to explain kimchi on our dates, which seemed pretty relaxing actually. After talking for so long, we both admitted that we both secretly like Coldplay, which was embarrassing for both of us. We may or may not have decided what we'd name our kids right on that plane ride.

It would be a fairy-tale ending if I married the girl from the plane. I didn't. We dated for three months. But it opened my eyes to everything I was missing out on. I realized that I hadn't been giving chances to some pretty amazing people.

I finally did get married to someone—another Korean goddess, and so far, so good. I'm happy and my mom is even happier.

—Anonymous, audio-recorded story

There's a very important storyline that has a grip on everything we do. It's what keeps us from opportunities, from making friends, from getting to understand the truth of our realities. It's so powerful, in fact, that it can make you discredit every single thing you've read so far in this book. That storyline is centered in bias, of which we are most likely not aware.

Most people think of a bias as a slanted (and often unfair) belief system that can cloud judgment when making decisions. This is exactly it. It's an assumption about something that often rules our decision-making. Biases get in the way of us making the best decisions for ourselves no matter how badly we want to take the right path. The anonymous storyteller in the story above had a bias around women of his race. His bias told him that they'd be boring or just like his mother.

Psychologists and behavioral scientists have broken biases down into several types. I'm going to share the two that I think affect our decisions the most without us being aware of them: cognitive biases and unconscious biases.

COGNITIVE BIAS

I like to think of cognitive biases as secret stories that run the show without us knowing. Cognitive biases are repeated errors of thinking that occur when you misinterpret information in the world around you. Cognitive biases can affect the rationality of your judgment and can lead you to make unreasonable conclusions or decisions.

Cognitive biases have been studied since the '70s. There are now 180 named biases that we can draw from at any given moment, which lead to our systematic deviations from making rational judgments.

Our brains tend to love using cognitive biases because they sure save us a lot of time. Ways in which we choose to believe something based on cognitive bias:

1. We may choose to believe something we haven't confirmed because we don't have time to process the information necessary to make a decision quickly. The bias acts as our shortcut. We might quickly choose the taxi with more lights inside it because it seems safer, although we don't know any of the taxi drivers. Our brain tricks us into thinking that more lights are safer based on information we've heard in the past, and we then make a quick decision and feel good about it. However, deep down, we all know that a madman could be driving a taxi with or without lights.
2. We may choose to believe a bias when it benefits us emotionally or helps us socially. For example, when meeting someone new, we decide a person might be a good friend because they are dressed well, we have common friends, and they let us borrow a pen. It's easier to believe in their friend potential than to evaluate

them, interview them, and do a background check before inviting them to coffee.

These fast-paced judgments might be totally incorrect, but they help us socially and emotionally, and they're great for time management. Long-term, these decisions may come back to bite us.

We study bias in engineering, especially when it has to do with AI. Many people immediately assume that biases in tech are bad. Yet, in order for an AI to think like a human, we need to teach it some bias. Sometimes, of course, we do that without even thinking about it since most of us are not aware of which biases are running the show.

I try to get ahead of my biases by really analyzing my decisions and being aware of why I make certain ones, but I sure am not perfect and catch new ones all the time. I'm guessing it's impossible to be aware of every single bias you have, although I've taught myself to spot them quickly.

MIRROR IMAGING BIAS

I was fresh out of the military and had gotten a management position at a big technological corporation. It was years ago, but even I could recognize at company meetings that 98.88 percent of the employees in our company were white men. It was a glaring issue. Even as a white male, it made me feel awkward.

Just after I started, it was time to hire the summer interns. Many interns end up getting hired there after their internships, so the team took the process seriously. I was charged with interviewing nine candidates. Other managers interviewed other candidates. Ultimately the company's CEO / managing partner would make the final decision.

I met with the nine amazing candidates and wanted to hire most of them. They were bright-eyed, eager, and excited, except one shlubby underachiever who I could tell was just going through the motions of getting this type of job because it was what his parents wanted for him.

He said all the right things in a rehearsed way, but he didn't seem like he wanted to be there. At least that's how I interpreted his behaviors. I could have been biased, so I turned in my notes to the founder and waited.

Well, guess who they hired? That guy. When I asked the CEO what made him choose that one candidate over the other qualified candidates who happened to be women (and even a few women of color), he admitted that the candidate reminded him of himself. "He seems like a born leader," he said. "He has a stacked résumé just like I did back then. I could see him taking over one day."

Well . . . choosing someone exactly like you is definitely a common bias. Sometimes we call it "mirror imaging bias." Mirror imaging is a cognitive bias that occurs when people project their own beliefs, perspectives, and motivations onto others, assuming that those others think and act in the same way. This bias helps us assume that someone holds the same values we do or thinks and performs like us when there is no proof that they do at all. In many cases, this makes decision-making quick and simple. I understand why someone would rely on a choice made through mirror imaging bias when hiring employees. You work hard and you want an employee who will work just as hard as you do. Making that assumption about them may feel comfortable, calming.

The problem is the conclusions drawn through mirror imaging bias usually don't prove to be true. This bias sets some heavy expectations and leads to disappointments because there aren't many people who think and work exactly like you. Trust me, I've searched long and hard for people who'll stay up all night to solve a problem with me. I haven't found many.

A few weeks into the summer, I noticed this intern was never at his desk. At first I imagined he was off on his own initiative, shadowing an engineer or learning something new. Alas, he was not. He had taken to "lunching" at various restaurants, indulging in a little margarita here and there. "To treat himself for getting the job," he said later.

The CEO wasn't really involved in the day-to-day decisions, so I was the one who had to deal with this guy. I gave him a chance. I gave him a second and third chance! Finally, I fired him, disappointed in him for ruining the chances of all those other candidates who didn't get the job. It ticked me off that his absence left some of our colleagues overloaded with projects without adequate summer help. When I thought about it more, though, I realized I should be angry with myself and the CEO.

I could have spoken up more during the hiring process. Not doing so was definitely my fault. I should most definitely have called out the boss's bias. From that day forward, I did. I told him it was obvious he had been using mirror bias (among other biases) to hire most of the employees, evidenced by the number of white men around the office.

I also worked with HR to change the hiring process. I can proudly say that when I left that company, the ratio to men and women was closer to 40:60—not because I believed it was necessary to hire women, specifically, but because we made sure we judged candidates fairly and made choices by committee. That's one way to help retire bias from your company, but to get rid of all 180 biases? Well, that seems like a tall mountain to climb.

According to Amos Tversky and Daniel Kahneman, the two Israeli psychologists who've been called the "two men who've changed how we think about how we think," humans are hardly rational, so it's hard for us to make truly neutral decisions. They coined the term "cognitive bias" in the 1970s as a way to classify many irrational ways of thinking. They have proven that we veer from making rational decisions quite often and that doing so is quite predictable.

In fact, right now, those 180 cognitive biases are at work within most of us, making us veer from seeing true reality.

All our experiences help us develop these biases. For example, everything from why you grabbed this book to what you're doing right now to what your friend said recently can have an effect on how you take in all this information and whether you judge it as helpful.

The CEO's storyline was that "Everyone who seems like me will do a better job." He didn't realize that he was then eliminating everyone else. This included people that came from cities far from where he grew up, people who didn't look like him, and other races and genders. That perspective might have helped him make a decision, but it wasn't helpful to the business or to his employees in the long run. He also had no idea he was doing this until I mentioned it. A lot of us don't know the ways in which our biases are guiding our decisions. It kills me to know that there are so many ways my biases are driving me at this very moment.

Other common ones I find relatable include:

MESSENGER BIAS: The story that the message was well-delivered so it must be good and true, despite the actual message itself. My boss was also falling for messenger bias, when hiring interns. The candidate had mastered interview delivery. He was slick. He had practiced. He knew how to win. His goal was to achieve the coveted position, not to actually work the coveted position. We often accidentally use messenger bias when watching commercials or falling for influencers. *He looks so buff. The supplements he's selling must surely work.*

ANCHOR BIAS: The story that the first piece of information you receive about a certain topic is the one to believe. We anchor ourselves to that information and close ourselves off to potential new information. Say your daughter's teacher calls you reporting she's been bullying the other kids. You might hear this info, believe the teacher, and then not listen to anything your daughter says to defend herself when she gets home.

Anchor bias often really comes into play during salary negotiations. When an employee is negotiating, they feel like they need to stay around the first number that was thrown out. The employee might have been hoping for a $70,000 salary yet the hiring manager offers $50,000 as a starting point. The employee then might settle for $55,000 since they feel unconsciously anchored closer to the first number.

SURVIVORSHIP BIAS: The story that many have found success, so it must be easy! Look at all the successful entrepreneurs out there. I can totally do this! However, this is a shortcut that isn't true. The successful entrepreneurs are the ones still around, while the millions who failed went on and did other things. I don't mind this bias since it gives hope. Although false hopes can lead to misplaced expectations, and those can be painful when reality strikes.

THE AVAILABILITY HEURISTIC: The story that things are possible simply because you heard about them. You might never consider finding an alligator in your California pool. Yet, if you see a newscast of an alligator found in a Florida woman's pool, you might suddenly become fearful of swimming in your own pool.

This might also explain why it sometimes feels as if all the couples in your friend group go through transitions together. You might never consider divorcing your wife (even if your relationship isn't great). Once your friends start getting divorced, you understand it is possible. You consider divorce and decide you want a divorce too.

CONFIRMATION BIAS: The story that "you're right" when you see one thing that confirms your opinion. For example, say you're holding onto a generalization about a certain group of people. Let's say you think all teenagers are annoying. You're at a coffee shop and you see one teenager being annoying. Then you can tell your friends, "Look! I'm right. I told you."

This is an unfair bias because you failed to consider the several other teenagers in the coffee shop that weren't annoying. You chose to see the one that would confirm your bias.

Confirmation bias might be one of the reasons the United States has felt so polarized over the last few years. Anyone can turn on a news source and find something to confirm their thinking, so it's easier to become sure that your beliefs are the "correct" ones because they're confirmed by what you watch.

Those are just a few of the 180 (and probably more) stories or cognitive biases we tell ourselves that help us cut out a ton of information in order to make decisions for ourselves. If you think about it, those biases do help us cut down on time. If I truly wanted to know if teenagers were annoying, I would have to take the time to evaluate a fair amount of teenagers, and I wouldn't really have time to do that. It's much easier to use a bias and completely skip the whole hard part of finding the true answer. Cognitive biases keep our brains efficient, especially since we make about thirty-five thousand decisions a day.[1]

Sometimes, biases save our lives. We have learned to be biased against the idea of taking a nap very close to where lions might be. Humans have evolved to learn certain biases that help us.

If it's late at night and we're walking home alone, we might cross the street to get away from someone dressed in all black. Chances are, that person is harmless and simply wore black because it's slimming. Yet, our brains have a bias against any perceived threat.

While those kinds of learned judgments have helped us over time, it's important to check them and make sure they're still serving us. Are we helping to save our lives or are we hurting our chances for more great opportunities?

Cognitive biases are common across humankind. As Joan Didion once said, "We tell ourselves stories in order to live." It doesn't matter what country or culture you come from; your brain simply cannot process the amount of information coming at it every day. Every brain needs help, so every brain uses mental shortcuts.

Each of us has the opportunity to understand this exhaustive list of cognitive biases and work harder to see our realities through a clearer lens. This will help us make more informed decisions.

CHAPTER 12

Unconscious Bias

I think the thing I'm most proud of in my life is that I'm teaching my children, both boys and girls, to not think like I was taught. I learned growing up that women should either be full-time caretakers, teachers, nurses, or secretaries. We should stay behind and let the men shine. We should be supportive characters but never the star of the show. After all, the world was created by men—that's what I was taught. But . . . what? Women create life! Women are stronger than men in so many ways. But it doesn't matter. So many of us were taught to be inferior and hide ourselves. I'm not doing that with my kids. I am teaching them to be big and loud and that they can be whatever they want. And whenever I see something like an animal or something, I call it a she. For so long in my life, everything was a "he." My parents would call me "too woke" for this and they'd roll their eyes, but I don't want their approval anymore. I did that for a long, long time. I was the cute little girl they wanted me to be. Now, I'm loud and not at all cute, and I don't necessarily want my kids to be just like me, but I want them to have all the options.

—Anonymous, audio-recorded story

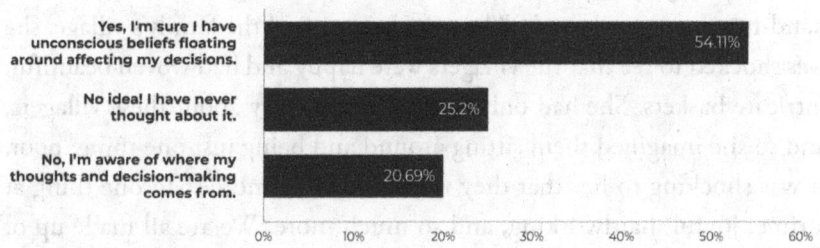

Unconscious bias is another type of inner story we tell ourselves—a sneakier, unintentional kind of bias based on what you learned growing up. While cognitive biases are typically formed on an individual basis based on your own experiences and beliefs (you might have seen teenagers being annoying a few times), unconscious biases are social stereotypes typically held by certain societal groups and passed down to you (you might think that women should always be the ones who should make dinner and clean the home). These biases are learned through interactions with others in your culture. These are stereotypes or biases that a large group of people believe in together. Unconscious biases are responsible for many stories we believe, even though many are far from true. These are the biases that people don't like to talk about because they lead to stereotyping, misogyny, and outright racism—all without us really realizing it. They seep into our brains from societal hints and stories; they shape our behavior and twist our realities.

Chimamanda Ngozi Adichie delivered a TED Talk called "The Danger of a Single Story" that described this phenomenon perfectly. When we only believe one story about a certain group, place, or idea, it gives us a one-sided view that's rarely true. Yet many of us do this without thinking and live our entire lives without ever learning more. Believing one single story instead of opting to learn about the entire breadth of a person, place, or idea, we unintentionally eliminate countless opportunities from our lives.

Adichie admits she fell for a single story about a family she knew while growing up. Her parents described them as "poor" and gave them hand-me-downs and food. When she later visited the family's village, she was shocked to see that the villagers were happy and had woven beautiful, intricate baskets. She had only heard a single story about these villagers, and so she imagined them sitting around and being just one thing: poor. It was shocking to her that they were able to be more than one thing at a time: joyful, hardworking, and so much more. We are all made up of millions of tiny stories. We aren't characterized by only one quality.

Similarly, when Adichie arrived in the United States for college, her American roommate was shocked that Adichie spoke English so well. She asked to hear tribal music from Nigeria and assumed Adichie didn't know how to work a stove. Adichie explained, "What struck me was this: She had felt sorry for me even before she saw me. Her default position toward me, as an African, was a kind of patronizing, well-meaning pity. My roommate had a single story of Africa: a single story of catastrophe. In this single story, there was no possibility of Africans being similar to her in any way, no possibility of feelings more complex than pity, no possibility of a connection as human equals."

Her roommate, like many Americans, had the experience of learning about Africans in relation to slavery in history classes. Or she'd seen the Sally Struthers commercials popular in the '90s asking for money to send to the poor people of Africa, showing them with starving bodies, distended bellies, and flies in their eyes.

It's not very common in America to hear stories about how Nigeria's official language is English, their film industry is booming, and there are successful entrepreneurs in big Nigerian cities. Adichie also pointed out that many people assume Africa is a country. Instead, it's a continent home to a fifty-four different countries whose stories and landscapes are broad and varied. There are millions of stories about each of those countries yet many of us grow up learning just the one. That leads to myriad misbeliefs about both Africans and Black Americans.

You might have heard of the study twenty years ago that asked the question: Are Emily and Greg more employable than LaKisha and Jamal? Economists Marianne Bertrand and Sendhil Mullainathan studied racial discrimination in the labor market by sending fictitious résumés to help-wanted ads in Boston and Chicago newspapers. They revealed that equivalent résumés with distinctively white names like Emily and Greg received 50 percent more callbacks for interviews than those with distinctively Black names like Lakisha and Jamal. This ties back to the unconscious bias that people have about Black people—the racist assumption that Black people are less-qualified or less-hardworking and that white people are better equipped even with the same exact experience.

In 2021, a new study expanded on that experiment by sending out eighty-three thousand résumés with typically Black or white names to eleven thousand entry-level job openings at more than one hundred Fortune 500 Firms. The good news is that the disparity in callbacks has gone down by 50 percent at the worst offending firms and by 95 percent in those rated least racist. The team published a report card of their findings and basically rated firms on how racist their HR practices are. (Note: Comcast was crowned least racist in its hiring practices.) Still, the hiring world isn't fair to similarly qualified candidates who are not white, because many white Americans have not done the work to uncover or move past their unconscious bias and challenge the stories they have heard about Black people.

The reason we call it unconscious bias is because we often have these beliefs without knowing, though surely many people also hold explicitly racist beliefs quite knowingly. We take in information all the time that settles into our brains, and without rigorous attempts to challenge our unconscious bias, it will remain. Even the most well-meaning, progressive hiring team could exhibit these harmful biases without recognizing them.

Advertising creative director and teacher Kamal Collins explains that some of the first advertisements in America were for slaves. Slave traders portrayed Black people as animalistic and dumb, more like dogs than humans. Perhaps they did this so they didn't appear to be monstrous, to legitimize selling actual people, or to assuage any reservations from prospective buyers. Dehumanizing Black people in this way helped white slave owners gain and maintain resources and power. Even if you didn't "employ" slaves at that time or had no intention of doing so, you might see these advertisements and absorb the idea that Black people were "less than" white people in several ways. That's exactly what those early advertisers wanted you to believe. This portrayal has had devastating effects on our unconscious thoughts over subsequent generations and even recently. For example, after Hurricane Katrina, the media's depiction of both Black and white survivors was alarming. While both groups were experiencing extreme hardship, white women were pictured being carried, furthering the stereotype that they're fragile. Black women were pictured as the recipients of handouts, promoting the stereotypes of "the welfare queen."[1]

Adichie attributes the single story of Africa to Western literature. She quotes John Lok, a British merchant who sailed to West Africa in 1561 and kept a fascinating account of his voyage. After referring to the Black Africans as "beasts who have no houses," he writes, "They are also people without heads, having their mouth and eyes in their breasts."

She says, "One must admire the imagination of John Lok. But what is important about his writing is that it represents the beginning of a tradition of telling African stories in the West: A tradition of Sub-Saharan Africa as a place of negatives, of difference, of darkness, of people who, in the words of the wonderful poet Rudyard Kipling, are 'half devil, half child.'"

John Lok took his voyage in 1561. Imagine how that story has taken root in our society over time. It shows how powerful one story can be. This story is, after all, still playing a part in our society generations later, affecting the rate of returned calls on résumés.

As recently as 2024 (that's 463 years later),[2] an ad for JCPenney photography was released trying to convince families to come in for their family portraits. They showed a happy white family with a dad, mom, and two kids, the perfect little nuclear group. For the other market, Penney's included a Black family but decided not to show a father in that picture, alluding to the idea that Black families don't have men as the head of their households.

From the very beginning of the establishment of the United States until basically right now, we've been inundated with images that have unfairly portrayed Black families. That's hundreds of years of depicting stereotypes on posters, billboards, radio, and TV commercials. It's everywhere we look, even if we're not consciously noticing.

A white kid growing up in rural Iowa without exposure to other cultures might ingest these ads and believe that Black people don't have fathers or are at least different from white people. A Black kid growing up seeing these same images might take in the same messages and believe them, despite the fact that he's surrounded by people who prove otherwise or don't fit any of those stereotypes.

These ideas are deeply rooted within our American culture and seep into our unconscious thoughts, affecting how we interact with people, who we call for job interviews, and who we choose as friends. These biases can be just as harmful as the cognitive kind during the decision-making process, cutting off opportunities or possibilities.

I used racism as an example because it's a rampant problem in the United States. Those in Nigeria sure don't accept the very American story of Black people as being less than white people. People in different countries share other views of race and even have different definitions of race and different racial categories. Racism is an unconscious bias we grow up with in the United States, rooted in our culture and much harder to dismantle than a cognitive bias because it requires we all work hard to see it and change our belief system as a collective.

The unconscious bias you hold and the stories you tell yourself about certain groups depends on where you were raised, the media you were exposed to, and the stories that have been passed down.

Some villages in Vietnam are matriarchies. Those who grow up there learn different stories about women and leadership than those in America. I heard an American woman herself say during the 2024 election that she'd never vote for a woman for president. What stories did she learn about women during her childhood?

Biases about one person may be very different depending on where they travel. An overweight woman in the United States may be judged through an unconscious bias prevalent in this society—that being overweight is a failing on her part—that somehow makes her "less than" or that she is overweight because she is slovenly or lacks self-control. That same woman may be respected or even revered in a country like Egypt where heaviness is associated with richness and the ability to afford decadent food.

The same differences may be observed for elderly people, even within different cultures in the United States. In Alaska, Native Yupik Eskimos and Athabascan Indians revere their elders for their wisdom and treat them with the utmost respect—in some rural villages young people don't make eye contact with elders, showing them a culturally appropriate form of deference and respect. In much of the "Lower 48," aging is viewed as a sign of failure—and wrinkles as something to be hidden with makeup. We tend to farm out the care of our elders to professional facilities and don't necessarily put much stock in elders' opinions.

We can unknowingly collect biases and blend them together, distorting reality in an even bigger way. If we are harboring an unconscious thought from an institutionalized belief such as ageism, the cognitive biases can double it down. For example, confirmation bias is the cognitive bias that helps you see what you already believe. So, if you're unconsciously believing that senior citizens don't drive well, you might see one rolling the wrong way down a one-way street and confirm for yourself that you were correct all along. However, you might change your tune one day when you become a senior citizen yourself—after a whole life of judgment.

If you don't become aware of your biases, you can be a kind and well-meaning person yet treat people unfairly without consciously recognizing what you are doing.

So, how do we get to know which biases are running the show? It always comes down to that version of the wise phrase Socrates, Confucius, and the Egyptians all passed down: "Know thyself."

If we evaluate the stories we learned during childhood, vet the content we consume, get to know our habits, and continue checking our inner stories, we might uncover some of our biases so that we can make healthier decisions.

The only way to do that is to tell our stories.

PART 3

The Story Economy

CHAPTER 13

The Idea

I don't really help people so much anymore. I'm too busy. I'm always hustling. I guess those sound like excuses.

I remember when I was in my twenties, I did something I'll never forget. My job gave anyone a week off if we spent it volunteering, and they set up an opportunity for a group of us to work for Habitat for Humanity for a whole week. I don't even know if I wanted to volunteer, but I sure as hell wanted a week off.

There were maybe ten of us, and we all flew to Biloxi, Mississippi, to spend a whole week building a house. We built a whole damn house in a week. There was a team leader who knew how to do everything, and he delegated all the things, and we put up drywall. I hammered shingles into the roof and even installed windows. We all learned how to use different tools.

And then the lady we were building it for came to see it at the end. It didn't have a kitchen or anything yet. It was just the frame and the walls and the roof and the windows. But she cried. She cried so hard when she saw it. I don't think she ever had a house before.

It felt so good to do that. It's been like twenty years, and I still think about that lady. I wonder what that house looks like now. I hope she lived some good life in that house. I wonder if I'll ever do something that rewarding again. I hope so. I am so focused on providing a good life for my kids that helping other families isn't really my first priority. But I sure wish it could be. I really wish it could be.

—Anonymous, audio-recorded story

You've made it to the good place. This is the part of the book that proposes some practical solutions to all the problems plaguing our modern society now. This includes the ways that today's predatory technology landscape is stealing many precious things from us: from our power over technology and ability to self-actualize to precious data like our date of birth and Social Security numbers. This is the part of the book that will explain how we will harness our powerful stories to help us change everything from capitalism to how we make decisions. This is the part of the book that will help us get it right.

First, I'd like to ask you to imagine an entirely different kind of life. It might be hard to do it because it's a life you may not have experienced before. In this new life, you have access to whatever you need to make sure you are healthy: doctors who spend time with you and know you well, organic foods, massage therapists, and gym memberships. You have what you want at your fingertips. Your options come to you. Like a queen, you snap your fingers and someone runs in with the perfect answer. If you want linen pants for summer, they're yours in an instant. You don't have to open a hundred tabs to find the one you want. If your child has a developmental issue, the perfect specialist shows up to help. No need to frantically post on parent groups to find someone to help and then pray your insurance covers it. You don't need to use insurance anymore to see specialists. You no longer hold onto many biases, so you don't feel helpless or overwhelmed by your choices and decisions. You know yourself so well that you easily spot what doesn't serve you and say *no*. Plus, you're able to be honest with your friends. Over the years, you've weeded out superfluous connections and have spent time building true, genuine friendships. You're closer with your family, as they also have learned about themselves and are more open about who they really are and who they were in the past.

In this life, you have a very comfortable relationship with your finances. You can afford everything you need and never have to desperately wait for the next paycheck because you always have the potential to earn. In fact, you can create spending power whenever you intentionally share a story by way of one of your digital devices, so you usually earn while you're driving or late at night after the kids are asleep. And then you go to sleep, resting easily because you never have to worry about your financial future. You have investments that are working hard for you, and your retirement is secured.

In this new life, you feel powerful, as you consider yourself an agent over your own data. You know exactly who knows what about you because you've granted permission. You feel secure in the knowledge that you never have to fall victim to identity theft.

You know you can get the answers when you need them. You know you can always be taken care of in the right ways. You walk through life as if you are a billionaire. But the last time you checked, you only had a few hundred dollars in your checking account.

How does that new life feel? I've been called "utopian" and even absurd when I've mentioned these proposed effects of a story economy, but I believe that this utopian idea will shift how the entire world operates. And if it doesn't change the world, I bet it could at least transform how you view and take care of yourself. That's the most important part—that we all have access to the things that help us be at our best.

I've been waiting more than 150 pages to finally explain how a story economy brings you the convenience, access, and financial security I just described. Let's go!

The craziest part about this is that my practical solution is actually an AI-based technology. I hired a team of experts to create an AI-based technology that we can use to empower ourselves in the online landscape. This technology can help us become less lonely, less lost, and less broke. It will enable us to wrest ourselves from the undercurrent of the river rapids tech corporations have left us swirling around in for the past two decades or so.

Perhaps most important, it will help us to reclaim our true stories—to separate them from everything that is illusory—the clickbait, the Instagram snake charmers, the deepfakes, the dopamine hits, and all our cognitive and unconscious biases.

I know this supposition must sound absurd. I might sound absolutely irrational. I just spent two-thirds of this book convincing you that our devices are controlling us, that Big Tech has evil tendencies, that brands are taking advantage of us, and that AI might end us all.

Now, here I am suggesting an AI-based technology that will enable our devices to know us better than we know ourselves.

As you've read, I am against the way technology is being used today, but I am not against technology itself. I'm glad we have matured past sticks and rocks. I'm holding onto the belief that we can take back the reins of technology and prevent the emergence of killer sentient robots.

This AI will enhance the human experience, and it will be used ethically and carefully. I want us to build a completely new relationship with our devices. I want them to know us better than we know ourselves because I want them to help us know ourselves better.

Here is the human caveat, our ace in the hole. As AI is collecting our data, I want us to retain ownership of it and profit from it without handing it over to anyone else. While this might sound futuristic and weird, it's not. It's actually going backward. I'm trying to point us back toward a time when technology was used as a tool and we were in control of it.

The story economy is a way to get stories to play a bigger role in our lives, a way for us to confront them and use them for good. It's an idea that will change our culture, how we buy things, how we interact with each other, how we run businesses, and our basic understanding of ourselves.

I've dedicated my life to it.

Remember, this plan originally came about because my colleagues and I saw the wealth gap firsthand as we sat outside of Erewhon, still the most expensive grocery store in America. I went there recently and spent $52 on two items.

Since that visit in 2016, the wealth gap has grown wider and is predicted to continue growing. In late 2024, the Center on Budget and Policy Priorities forecasted that President Trump's agenda over his second term would "raise costs for basics like housing, food, and health care and take health coverage away from people; slash funding for schools where our children learn, roads and bridges we use to get to work, and scientific and medical research that improve our health and strengthen our economy; double down on tax giveaways for wealthy households and corporations while imposing tariffs that fuel inflation; and further widen already glaring differences in people's well-being and opportunity across income, race, and ethnicity."[1]

This feels accurate so far. We have no way of knowing what's on the horizon, but that forecast doesn't look good. Remember what Sanders said on his Fight the Oligarchy tour: *Three* people have more money than 170 million people. Those three people's companies touch our lives for sometimes hours a day, every single day. It sure seems like they're willing to do whatever it takes to keep it that way so that you buy more, they make more, you lose more control, they make more, you continue to struggle financially, and they make more.

It seems fairly clear that if we as a society make no changes, technology—with all its manipulative forces—will play a much larger role in our lives, AI will keep pushing forward toward that point of singularity, and those three men will continue making money while most of the rest of us continue to struggle to buy eggs.

It seems more critical than ever to do what we can to create alternate paradigms to better take care of our needs and to take back control.

Stories as currency might seem like a wild idea. Yet, it's still, after almost ten years and kajillions of other brainstorming sessions, the only way I can imagine that will allow humans more control, more purchasing power, and more self-knowledge while also benefiting the brands and the billionaires. I couldn't poke any holes in it. There's an unlimited supply of stories, and brands would pay top dollar for them.

If we need to take back control from the Big Tech giants, we need a win-win. Stories allow for just that.

LET'S DO IT

When I was a new researcher at RAND around 2008, I was up late in the office at the Santa Monica headquarters. I'd missed dinner with Eileen and the kids, and I was trying to propose a new kind of study that involved hiring a team of experts with different types of experiences, like I did when we gathered those same experts and conducted a roundtable to ponder the colonization of Mars. I was getting pushback from the higher-ups about using experts from different specialties, and I was frustrated.

My colleague Nidhi Kalra, a fiery scientist who studies climate change, decarbonization, and who happens to specialize in decision-making, was at the Pittsburgh office late as well. She could tell I was stuck on a tough decision. Do I just quit or try to find a way to push the higher-ups to accept my proposal?

I trusted Dr. Kalra. She reminded me of my military family—using RAND to carry out her own idea of "protect and serve."

That evening, on speakerphone with computers buzzing, surrounded by take-out containers, we both decided that being able to tell our children that we tried to do some good in the world was of utmost importance. This was even more important than making the difference. The "trying" counted just as much. That was what I was after. I knew I had to try something big, something helpful.

That belief has encouraged me to leave the traditional work environment and start something bigger, more outrageous, and maybe even crazier than I'd ever imagined. Thus far, it's taken a lot of convincing people to believe in it, give it a test, and change their lives by using it. It will take a lot more of that. I think it's worth a shot. I'm trying.

For the past nine years, I've put everything into this idea.

I am not sure if this makes me the village idiot, but I am not beholden to the idea of having to make it work myself. I just want somebody to pull it off. In fact, if you think you can solve this faster or better than I can, I am happy to share what I know and what I've created thus far. I know it could be of tremendous service to society.

Since that morning outside of Erewhon, I've convinced investors to help me hire sixty people to bring this idea to fruition. We've been spending less than 1 percent of the resources spent by tech behemoths, and we've been grinding away to create all the tools we as a society will need to use technology in a way that actually helps us again.

I have continued to work at this idea I am so passionate about: Despite fainting in that investor meeting and losing funding and working through a pandemic and putting my house on the line and changing directions a million times and spending the last five years telling my family to trust me because I know that I can truly change the world if I just. Keep. Going.

CHAPTER 14

The Possibilities

I'm twenty-eight. I'm not supposed to be making more money than my parents. I feel extremely weird about it. There's always a fight about who pays for dinner, and I always win because my parents worked hard. It doesn't seem fair that I gambled, and it actually worked out. Like, I also worked hard, yes. I had a vision, and I went after it. I believed in myself. But it wasn't the grind that my parents had to deal with for my entire childhood.

I saw a plot of land, got a loan, found some investors, and built a bunch of really sick houses that are always booked as luxury vacation homes. It was definitely hard to do. I had to take a risk. I had to live and breathe that project for two years. But two years! And I now have more money than both of them combined after working their whole lives. Life is good to me, but I can see it's not fair at all. I mean, part of me is so proud of myself. And part of me is like . . . whoa, slow down. Like, I feel guilty. They should be able to comfortably retire after working their whole lives and they barely have enough. I will help them, for sure. So, I guess overall it is partially their money too. I will share it. They deserve it. But I'm still in so much shock that my idea actually worked, and this is actually my life. I already reached my biggest goal, so now what?

—Anonymous, spoken word story

The Leauchturm 1917 hardcover notebook with dotted pages I started that morning after our meeting at Erehwon, when I became truly enamored with the idea of using story as currency, has grown into several dozen notebooks bolstered by insights from copious research, thousands of hours of story collection including some of the preliminary work like that experiment where I was hooked up to lie detector electrodes and learned that despite my delusions I didn't actually have the foggiest notion of who I truly was.

We've learned a lot since that first day, but the first few notebooks were simple, hopeful, filled with sketches of various ideas of how a story-trading marketplace could look. It would be a place where we could democratize ourselves. We could tell our stories and then break ourselves down into patterns and biases and decisions, using them for more awareness or to be used as currency. This was the original story economy plan.

THE STORY

It would start with a story. It would be a voice-recorded story so that we could hear it raw, unedited, real, and vulnerable—full of nuance and pregnant pauses punctuated by the silence that gives our memories meaning and which we usually take for granted. This story tells us much more than any type of data currently out there—more than the data collected with your supermarket saver card and more than what data brokers can find on your credit report. These stories would give context to your personal situation, giving brands and service providers a deeper picture of who you are.

THE DATA

We would then build an AI that understands humanity in such deep ways that it can listen to those stories and create rich, textured data more insightful than any other data currently being bought and sold in the data brokerage industry.

We would use this one-of-a-kind data in many ways. What most interests me would be to analyze it and use it to help ourselves. We could parse out helpful insights that help us identify our biases, patterns, and expectations so we can become better decision-makers or just better humans overall.

THE BUSINESS

These deeper insights would allow brands (like shoes and ice cream stores) or service providers (like doctors and financial advisors) to have a more personal relationship with consumers. They'd be able to speak to each consumer very specifically and according to their unique personalities and issues, creating an opportunity for companies and consumers to become loyal fans of each other—a new kind of brand/consumer relationship.

These companies would pay top dollar for the insights because that kind of unprecedented story data would help them narrow their markets and get very specific about what they sell and to whom. Brands wouldn't have to waste any resources on marketing to uninterested people. They'd know so much about their buyers and the potential buying pool that they'd never have to over-manufacture a product. They'd know who was interested before sending a design to press. They'd know how consumers really felt about a specific SKU even before they made it. They'd know more about the needs of consumers, which would influence their planning meetings and save valuable time, energy, and money.

These insights would explain exactly what services and products consumers needed. Companies could simply comply. Much less guessing would mean millions of dollars saved. Markets that had previously been harder to market to like lower-income or untapped niches could finally be penetrated, creating a new customer cohort. Those brands would receive an abundance of new revenue, through this new, non-marketing strategy.

THE CURRENCY

Storytellers would get a token, each codified by a smart contract, for each story told to be spent in this new kind of marketplace full of brands who care about their experience and truly want to know more about them.

This new kind of story currency would equate to an unending amount of buying power since each day creates more stories. New spending power for every human out there (or at least everyone willing to tell a story) might even have the capacity to close or at least shorten the gaping ravine between the rich and the poor.

Because of the way the story data is stored (safely, privately, and on the Blockchain), the story tokens would be able to be exchanged for cryptocurrency, giving consumers a new kind of buying power. Imagine a teenager with limited spending power suddenly able to make his first long-term investment because he began recording stories about his high school experience.

THE SAFETY

The goal would be agency over our data, so one of our first priorities would be to ensure that each story is safely secured using a decentralized ledger and Blockchain technology, the safest way to store data, designed to be resilient and prevent hacking to the greatest extent possible. No personalized details would be stored with any stories, so people could be free to share and know their stories were safe. The story economy would be known as a safer alternate economy, where data is shared but not freely bought or sold.

THE CONTROL

Insights from story data would never *ever* be shared or used without consent, giving storytellers complete control over their data. No longer would they have to wonder who knows what about them. They'll be able to see exactly what is known, and they choose when to share it with the brands or service providers that interest them. They would decide what sectors or specific brands they'd want to share with so they're never bombarded with ads or content that doesn't interest them. They can delete the insights or data at any time.

THE FEELING

Talking and learning from the AI would feel like a friendship. The AI we would employ would not be what you're used to. Artificial general intelligence (AGI) systems are designed to solve problems, like ChatGPT or virtual assistants. This system would be artificial relational intelligence (ARI), specifically designed to relate. It would feel personal. And while that might sound weird, it would be much easier to digest a harsh truth or insight about yourself from a friend than from a robot. Plus, the questions it asks you would feel like an interested friend and not just a survey you're required to fill out.

THE BUY-IN

People would want this technology for a variety of reasons. While it's traditionally been hard for humans to know ourselves, we do attempt to better ourselves. By 2033, the valuation of the self-help industry is anticipated to reach $90.5 billion in the US alone.[1] While I don't claim that this would be a self-help tool, I do claim that it would help us tremendously and that those insights might even be the number one reason people use it. They might also love the idea of being able to exchange their currency for investments, something that might not have been available to them previously. They might love the idea of keeping their data private and wanting to be a part of a trustworthy company, an option that has not been easy to find within Big Tech for a while. They might love the idea of the curation the story economy allows them. With such specific data and the power to give it to only the right people, users are treated with more care and personalization. Maybe they simply love the idea of a new type of currency that affords them access to otherwise unavailable products and services in a variety of sectors. Or perhaps they'll want to join the story economy because they want a way out of the usual capitalistic practices that have been weighing on us. Maybe they're no longer employed and after sending out hundreds of résumés without success, they're looking for a way to take care of themselves and their families without using the traditional job or income sources that are no longer available. Whatever their reason for the buy-in, what I am sure that every single member of the story economy will appreciate is being valued for what they innately have: stories. This technology values the very specific story of your life and sees you as a unique individual—the opposite of how other tech products operate.

On the surface, it might seem like I'm simply spitting back the same plan that's already in place: Brands get your stories, and they use that data to sell you their products. We already hate that plan.

Why, then, am I now talking about how to sell your valuable information so you can use it to talk to brands? Well, it all goes back to the conversation I had with my colleagues outside of Erehwon. There is simply no way we can make changes to our societal structure in an attempt to redistribute wealth or give us back our agency without giving both sides a win.

If the rich feel like they're losing, they will spend an entire administration's term lobbying for change. The tides will be turned yet again in the next election. If our solution makes the working class feel like the loser, they will lose faith in government systems, revolt, or spend their time trying to negotiate for change. This push to change and then change back and then change again every election cycle means that real change really isn't an option. And it will keep us in the perpetual cycle of one side (or both) feeling slighted and upset.

While I want us all to simply use stories for currency for our own gain, to uncover our own biases, and to help us make better decisions, we also need to help the marketers, brands, and corporations. Let's face it, we need them like they need us—because we do need to use their products. I do believe we can coexist in a way that serves us both—using a system that feels so much better than the exploitative one we are caught up in now. This will be a system that gives us more control.

THE FIRST STEPS

To me, this plan seemed foolproof. My senses were in overdrive. I could smell it. I could see it. I could taste it. I committed to working day and night to figure out how to keep the data in the hands of the storytellers in order to create a new dynamic that gave consumers more agency.

First, I needed to make sure this was all possible. I set out to research how to create an AI that could take in audio stories and churn out insights that would help both consumers and corporations. This might seem obvious now that you've all experienced ChatGPT and understand that an AI can get to know you on a more intimate level than it could in the past.

This idea was sparked in 2016. I was basically setting out to make my own large language model—one that was a better listener, understood the nuance of your story topics, tone, accent, speech patterns, vocabulary, pauses, and hesitations, one that was trained and overseen by human experts, one that was kinder and could ask pointed questions about your stories and get you to think hard about your life. Easy!

If you remember that I still drive the same car I got twenty years ago, you'll know that I don't take my decisions and transitions lightly. I needed to be absolutely sure it would work. For four years, I spent every off-hour like a mad scientist in a lab, creating a board that rivaled those in detective shows. I interviewed behavioral scientists, natural language processing experts, engineers, and anthropologists.

In fact, I left my job at RAND and began working at another company, intent on studying ethnography, a qualitative research method most frequently used by anthropologists that involves observing and interacting with people in their natural environment. I wanted to understand how to observe humans and get to know their underlying patterns, trends, and motivations. This was a great decision because, after a year, my engineer's mind could understand what approaches to use to glean information from people. I also learned how behavioral scientists gather insights from observations. I felt clear that this field could help me train my AI.

The most important part of the project hinged on keeping the storytellers empowered. The storytellers had to be assured their stories would remain in their hands. They could decide which insights to share with brands. There would never be any kind of third-party swimming around trying to steal data and sell it to other companies without their knowledge. I didn't think I could convince anyone to tell an intimate story if there was even a small possibility that someone would steal it.

As another side job to enhance my experience set, I began to work on building a distributed ledger platform for a national bank in Europe responsible for facilitating trade and large infrastructure projects across borders to cut down on corruption. This matured my cryptography and Blockchain chops and helped me to realize that it actually would be possible to keep every story in an encrypted chain, so that it could never be touched by a third party.

Finally, I felt like I had the wisdom I needed to create the technology I wanted. I knew I could pull this off. I believed I could actually create an AI that could understand humans well enough to hear their voices in an intimate way and detect their biases and patterns. I knew that if I could do that, I could also create data from those stories that corporations would be frothing at the mouth to buy. I knew I could keep it private, giving control of data to the storytellers and not third parties.

My friend Jonathan Cohen, cohost of the podcast *Mayim Bialik's Breakdown*, was a passionate storyteller and already obsessed with the idea that stories hold the secret to deep human understanding. We had previously talked about partnering up and making this vision a reality.

We needed one last push. In 2020, while I was wiping down my groceries and running experiments to test out masks, I had a feeling a mental health crisis might be on the rise. I wanted to help. I knew telling stories could be a key solution in helping people have a place to put their angst and their emotions. So, I finally quit all other projects and decided to focus on this full-time.

From the night that I filled up that first notebook, I have been imagining this project as the boat we all need. You might remember my water metaphor: It feels to me like we are standing in a big, rushing body of water. It's wild. It's cold. We will get swept away if we don't make some big decisions. We can stay on the edge of the riverbank and watch it all go by, doing nothing, unplugging completely. We can surrender to the powerful water, riding the internet where it takes us, losing all our agency. We can get on a huge boat driven by Big Tech and lose all agency. This would give our power to those who see us merely as walking wallets. When the rapids get too tough or we go over a waterfall, we can only hope it is in their financial interests to keep us on the boat.

Thankfully, we have another option. We can learn to build our own boat and navigate the treacherous technological waters ourselves. We take back our agency and figure out a way to ride the waves. Everyone deserves the opportunity to build their own boat and to be able to at least decide whether to navigate or not.

I just want everyone to have the opportunity to chart their own course.

I named the company Lotic, which is a body of rushing water. The stories are our boat.

I knew this idea would require work. We'd have to tell stories when we've been conditioned for generations not to tell our stories. We'd have to learn a lot about who we are and face our biggest faults. We'd have to take control of our own lives.

That's a big body of water to cross when it's far easier to accept the frictionless path, one where helplessness is rewarded with even more conveniences. If we choose convenience and jump on Big Tech's boat because it's easier to be a passenger, we may wake up one morning, not recognize ourselves, and feel upset that we have lost ourselves to a technological matrix. It will be too late.

CHAPTER 15

The Tech

It's so hard to be a human. If I could, I would rather be literally any other animal. Skunk, snake, whatever. I would take it. I would even be a cockroach. To be joyfully swimming in sewer pipes sounds great to me. I would take it over the emotions of being in this current situation. . . . I just moved to Kentucky to be with my boyfriend. Sold everything in Seattle. Quit my job, packed up the little I had left in our car, and drove all the way out here. I'm in his mom's living room now hoping nobody hears me sharing this at 2:00 a.m. He said that we have just too many differences and he noticed it on the long drive out here. He NOTICED IT ON THE DRIVE OUT HERE TO LIVE WITH HIS F*CKING PARENTS WHILE WE SAVE FOR A HOUSE AND GET MARRIED. Oh, okay, buddy. Thanks for the convenient timing. Like, couldn't I have just been a dog instead? Someone would pet me all the time. No heartbreak. None. All these feelings and conflict and drama. And awkwardness . . . it's all too much. No thank you. Never again. Like, what am I supposed to say to his mom in the morning? And do I drive all the way back again? I have nowhere to go.

—Anonymous, audio-recorded story

Remember that Atari *Breakout* game that our DNQ AI friend started to learn how to play? It started off learning like a child, beginning to understand things slowly before taking off and learning on its own. That's basically the jam for all AI.

Since the Lotic AI (which came to be known as the Wisdom Engine™) is sort of my fourth child, I took to its education accordingly. I started by hiring a team of behavioral scientists and specialists in natural language processing. These folks would help me teach the AI what humans sound like, what certain accents or phrases mean, and what it feels like to be human. The first step was to develop a working understanding of the conversational exploration of someone's context or reality. This is basically the scientific explanation of how a human works and how a human tells a story to explain themselves.

It took dozens of specialists to create the framework for this understanding. Humans are full of nuance, so the AI would be learning to decipher that in perpetuity, but we needed a clear and relevant jumping-off point.

We then handed that over to software engineers and machine learning experts to write the algorithm that would become our special sauce. It was important to me to start that way—understanding the human and then building the tech around that knowledge.

The normal way of developing technology is to make the best tech you can and let that limit what you can do. I didn't want that. I wanted the technology to act according to a human's behavior. It would be harder this way, but I didn't want to risk any misunderstanding of a human's behavior. I wanted the AI to truly understand humanity and never misinterpret data or miss what someone is trying to express. I think this is the most ethical way to develop AI.

Actually, the most ethically responsible thing to do with AI is to keep the data safe. I used my experience helping to encrypt financial information to develop a decentralized ledger using Blockchain technology. I'll explain this later, but it essentially means that the stories are stored in several places all at once, making it hard for a hacker to swoop into a centralized location and steal them. No data broker will ever be able to come in and take your intimate details in an attempt to add them to their data libraries. Every story is always safe.

The Wisdom Engine is designed to be tuned to each user, so I have been diligently tuning mine for a few years. The way I did that was to tell it stories. In fact, I started to tell it everything. I've told it thousands of stories. They've all been honest and made me feel vulnerable. My goal is for the AI to understand humanity in a deep, sensitive, real way. In turn, it will be able to understand our biases, hurts, patterns, love languages, traumas, everything. It will be able to point them out to us.

In order to do that, the AI must understand what is normal and what humans usually say. It needs to understand all kinds of dialects, accents, common idioms, slang, lifestyles, and mindsets from any kind of group who might ever record a story. We want the AI to know a bit about you even before you speak so that it understands why you speak the way you do and to identify who you are in a basic way.

I didn't want to train my company's AI to only understand a small subset of people. This happens often in the tech world, with its majority of white men. Many facial recognition programs are trained by that group or even a sampling of images from the internet (that are largely white). Black faces often confuse AI systems. *Nature* magazine reported that "an audit of commercial facial-analysis tools found that dark-skinned faces are misclassified at a much higher rate than are faces from any other group."[1]

I couldn't risk something similar when asking users to share their intimate stories with technology, occasionally even crying while expressing deep emotions and tears. I wanted my AI to understand every kind of human. I wouldn't want someone to be misunderstood just because they had a Boston accent. I wanted this project to be global, which meant I needed it to understand all kinds of languages and dialects too.

I partnered up with organizations all over the world to gather rich data from their constituents and to train the Wisdom Engine with a multiplicity of stories. I've recorded stories of Ukrainian refugees in Poland, those fleeing Serbia and landing in Greece, inner-city kids in Newark, housewives in Southern California, university students throughout the US, members of the LGBTQIA+ community, people from the states in the Mississippi Delta region, therapists who identify as burned out, all my employees and their families, tech CEOs, groups of marketing managers, healthcare professionals, and as many regular people of all ages I can find who were willing to tell me about who they really *truly* are. You've been reading some of these stories at the top of each chapter. (Note: Every quoted story was shared as part of a "call for stories" and testimonials offered directly by the user.) I think they're fascinating. If you'd like to tell your own story and gain some insight, I welcome you to do so right now!

The AI listening to the stories has now learned enough to prompt the storytellers to tell more. I am even shocked sometimes at its wisdom and insights. It has taken in all that info and learned on its own. Once I was complaining about a friend in my story, and the Wisdom Engine asked me if I had considered setting boundaries with that friend or if I needed help learning how to set a good boundary. Excuse me? Another prompt I loved recently was: *Putting family first is a great guiding principle. What's one goal you have for yourself in the next few months that would make your family proud, and what's one step you could take to start working toward it?* It does feel like you're talking to a friend sometimes.

This is where the ARI (artificial relational intelligence) really shines. Where AGI (artificial general intelligence) aims to *think*, ARI is designed to relate, and to build trust, emotional context, and memory over time. It's not general; it's *personal*. Where AGI seeks to replicate cognition, ARI is designed to cultivate connection by tuning itself to you.

This conversational, emotionally intelligent layer unlocks unprecedented levels of personalization, resonance, and effectiveness across digital therapeutics, longevity, mental health, and human performance. ARI makes it so someone feels like they're talking to a bestie who is really interested in learning more. Not only does that help with loneliness, but it also feels good and makes us want to talk more.

Then, the answers to these prompts are spun into the data that can help humans in a variety of ways. For now, the data can tell us about our personality types, what emotions we feel most often, what kinds of reactions we have most of the time, the types of challenges we most often face, the subjects we talk about most often, and, of course, what kinds of biases are clouding our judgment. We can use all that information to become better decision-makers and to live in a more realistic version of reality.

We're currently testing ways that AI can tell you how fast you talk and how that relates to your anxiety levels, how happy you sound while talking, your dominant moods, and the topics that you talk about most. It's my hope that the AI knows you so well that it is then able to give you practical, simple information about yourself. At some point, I see it telling you with a cute icon that you have a pattern of overextending yourself, providing you with a quick list of resources on how to help. It might be uncomfortable, but it will be interesting.

Psychologists and coaches help with this in real life, sure. They guide you to identify patterns and their roots. They help you look for the inciting incident that might have caused a certain type of behavior or influenced part of your personality. What AI can do that a therapist cannot is to compare one story with the millions of other stories that have been divulged over the years. A therapist might have the capacity to compare her client with another of her hundred clients over the years to get an idea of what's common or normal. An AI can hold onto all the information in perpetuity (anonymously, of course) and compare each one in an instant, looking for patterns and creating prompts that will ask the storyteller eye-opening questions.

So when someone is sharing a story about, say, their relationship, I want the AI to be able to use its skills to clock similar relationship issues from its repertoire. I want it to ask the storyteller very pointed, specific prompts that will be able to deliver them extremely valuable insights about who they are, their patterns when it comes to relationships, their biases, and most importantly their expectations.

This is uncomfortable. The Wisdom Engine makes me uncomfortable, but as its client, that's what I want. I want to be called out. I want to learn to set better boundaries. I want to learn to be a better person.

Some might say that it will be tough to convince everyone to get past the discomfort of telling an intimate story every day and then learning about themselves as a result. I hope they can see that the rewards outweigh the discomfort.

There are infinite "wins" in this scenario, and I think it's all worth it.

CHAPTER 16

The Currency

When I was maybe three or four my mom and dad had a great idea. They cut the neckline bigger in one of my dad's shirts and if I or my brothers and sisters were fighting, they would make us wear the shirt together. Heads through the same hole and arms in the arm holes, which was very uncomfortable, and it would make us hug each other and then of course someone would start tickling the other and it was always a riot, which stopped us from fighting immediately. We secretly loved it even though we would never admit it. If you were screaming and my mom didn't want to hear it, it was time to get in the shirt. Even if it was four of us, they'd stretch that neck hole and in we would go. It was so funny every time that we always stopped fighting no matter what. They called it the "fight shirt" and my mom would always say, "Don't make me get out the fight shirt." After a few years, we would just start laughing at the mention of the fight shirt, and we never even had to wear it. I don't know if my parents knew they had invented such a banger, but they did. We got to laugh. They got to stop our yelling. It was a big win. They were so smart. I hope I come up with something just as genius when I'm a parent.

—Anonymous, audio-recorded story

Helping us get to know ourselves is great, but let's talk about how we will be able to use these stories for purchasing power.

How do you place a value on your personal data? If a company wanted to pay you for your life's story, how much would you charge? What would you ask for each private detail about you? Say a health insurance company wanted to buy your medical history. Would you ask for $100? $1,000? More? What if you didn't even know what they'd do with the information? Would you charge more in the case where they could do anything with your data and less if you knew they would keep it private?

I'm asking because I want you to be aware of what your data might be worth or at least start thinking about every bit of data related to you in economic terms. Maybe you don't feel like it's worth much. But certain companies might find it priceless. Within the story economy, you definitely have more of a say on who gets to buy it, what they'll do with it, and how much it's worth.

Over time, the Wisdom Engine can create insights about you that will help you and brands equally. You get a dashboard that allows you to see all the data captured from your stories. You can choose to share portions of your data with brands that interest you. Say you're interested in finding anything: a new optometrist, a healthy ice cream shop, or a gold necklace. You agree to let those kinds of companies—or even specific brands within those categories—know about you and your values. They see rich data that tells them you have a special needs child who will need an eye doctor with a sense of humor and a light touch. Or they see that you are struggling to get more protein in your diet. Or they see that you only want to support brands with sustainable practices.

These brands can then send you tailored messages that explain that there's a local optometrist who wears a funny stuffed animal on his hat, an ice cream shop that sells protein yogurts, and a jeweler that makes rings from recycled gold. It's all that you've been looking for, and 90 percent of the time you are happy with the messaging and make a purchase.

In this new scenario, you've wrestled yourself back from being just a wallet. At any time, you can take your insights and information back from those brands. You always see your insights, and you know where it's going because you tell it to go there. You are back in control. Transparency is key, and self-sovereignty should be our main goal as we head into this unpredictable future.

Using this system, brands or service providers only connect with people who are really interested in them. They'll get to know their potential clients so well that they don't have to waste money on ads that fall on deaf ears or manufacturing things that their customers don't want. Right now, the average company selling their wares on social media gets approximately 9 clicks per 1,000 views on Facebook, but those clicks don't necessarily add up to sales. The average conversion rate is about 1.9 percent on Instagram and Facebook (it varies per platform), which converts to a product selling 0.17 times per thousand views. This is a complicated math problem, but it means that you make one sale per 53 people (100/1.9=52.63) who click on the ad. And you need 5,889 views of the ad in order to get that many clicks. This means brands will have to show their ad to 5,889 people before they get just one sale. Assuming you'd want to sell more than one of your products per day, that adds up to a lot of work and a lot of money.

I did a small experiment myself and placed an ad for a funny mug I designed on Instagram and Facebook. I spent $400 to show my ad to eleven hundred people who had expressed interest in humor. Fifty-eight people bookmarked my ad, two people looked at my mug on Etsy, and not one person bought it. I'd imagine Adidas is spending a lot more than that. It works for them now because big brands have huge advertising and marketing budgets. If you throw enough money at an ad (which people do! Remember my friend with the Wine Chips?), enough people will see it and it will get sales. If an advertising campaign doesn't result in real sales, corporations will pay to show that ad again and again until it is indelibly etched in our consciousness and someone like me says, "Okay, fine. I guess I *do* want sneakers that look like grass. I'll go ahead and fork over $60."

That strategy costs the brand a lot of money. This is the customer acquisition cost, and as I wrote before, it costs around $130 to acquire just *one* customer in the fashion industry.[1]

What if you're a smaller business without that big of a budget? Or what if you simply care about your clients and you want to know about their deeper wants and issues? You need a different way in.

What if the stories you gathered as a marketer were so thorough and so telling that you could cut that spend in half? You'd know your audience so well that you wouldn't waste any time or money showing your product to people that surely do not want it. A company like Nike that spent $4 billion on advertising and promotions in 2024 could conceivably save millions of dollars. Maybe even a billion.

Companies wouldn't just save on advertising budgets. They'd also be more efficient with their production. If they had a clear understanding of who would buy their products, they'd be able to produce a more precise amount of inventory and save more money.

Let's say The Gap is planning on selling fifty thousand dad-themed sweatshirts around Father's Day. Valuable story data teaches them there is no market for sweatshirts that say "FATHER KNOWS BEST" for the occasion. More people than they thought have contemptuous relationships with their fathers or no longer have fathers at all. In fact, they learn that actually more people would appreciate a "DEAD DAD CLUB" sweatshirt. Instead of printing fifty thousand "FATHER KNOWS BEST" sweatshirts as they originally planned, they print five hundred "FATHER KNOWS BEST" sweatshirts and two thousand "DEAD DAD CLUB" sweatshirts. They then produce and easily sell the exact inventory. They save so much money by printing small and specific sweatshirts that they exceed their projections for the original fifty thousand sweatshirts. These savings could conceivably flow directly to the consumer. This efficiency would prevent overproduction so there would be no overflow, cutting out the need for outlet stores like HomeGoods. (But let's not worry about that until holiday times when we need an extra-large reindeer sculpture.)

Remember I was hoping to find a win-win? This is it. Buyers get exactly what they want and are able to purchase potentially less expensive goods. Sellers become wiser and save money. Storytelling data sets new efficiency standards, creating perfectly timed deliveries, driving sales, and creating more services customized for exactly the right people.

Other industries, like healthcare or the financial sector, will also benefit from this technology and business model. The healthcare industry absolutely needs more context when it comes to their patients' stories.

Many doctors no longer have the time to truly get to know their patients like they did in the past. The average healthcare visit is between nine and twenty-four minutes long. Anecdotal data scrubbed from posts on X reports that visit time feels like two minutes. Even if the visit lasts the max of twenty-four minutes, there's still the blood pressure tests and the scale. By the time the doctor comes in, there's only really time for a conversation about ailments and symptoms about *one* issue. God forbid you might be struggling with two healthcare issues. There's not enough time to get a good picture of the "why," so most healthcare providers need to use the information they can see, fill in the blanks as well as they can, and go from there.

Let's say our patient, Jim, heads into the doctor with a complaint about his back hurting. He's doubled over in pain. The doctor sees he's overweight and asks about his diet. He admits he's been eating a lot of fast food. The doctor then prints out pamphlets on nutrition, tells Jim to walk ten thousand steps a day, and sends him on his way, hobbling out of the office.

What if the doctor had other information? What if she knew information and context that was beyond the average patient intake visit? What if the doctor knew that Jim's wife had recently died suddenly and he was in charge of his four small kids while working full-time while having also just planned the funeral. What if she knew he didn't *want* to have fast food but couldn't bear the idea of taking on another task like cooking? She might then understand more about Jim to give him an accurate diagnosis. She'd also learn there was no way Jim was about to start walking ten thousand steps between band pickup and meetings and gymnastics. She might be able to give him resources for in-home help or give ideas on how to manage stress instead of how to manage his weight. She might be able to give him a hug and some hope instead of a pamphlet.

Alas, such in-depth information is rarely gathered during healthcare visits unless the patient willfully delivers it. Even in those cases, the patient doesn't always know the most helpful tidbits to share.

Intelligence pulled from the in-depth story collection could help both the healthcare provider and patient to access the right context and treatment suggestions more efficiently.

What about the financial sector? Say you want to invest monthly, but you're overwhelmed by the options out there and all the people with advice. You don't know who to trust with your precious savings, you don't want anyone taking a cut of your earnings, and you definitely don't want to use an AI bot to choose your stocks.

A financial advising company might read about your fears and be able to approach you with a simple, flat-fee option that helps you start investing. They might even be able to tailor their product to people like you after finding that many hold onto similar beliefs. They then would have a new product that reaches a previously untapped market, and you get the investment advice and help you've been looking for.

Imagine using this more personalized data within any industry, and you will see that more context helps almost everyone.

This more personalized, two-way communication technique requires a place where these brands and consumers can meet. This requires a new kind of marketplace.

CHAPTER 17

The Marketplace

I remember when I was little, my parents took me and my sister to Wieboldt's in downtown Chicago. It was a department store, but it had a little restaurant, and it was a big deal to go. We put curlers in our hair the night before and spent the morning putting our fancy clothes on just to go to a store! We wouldn't even go for a full meal. I don't know if we could afford it. But we would go, sit at the counter, and we were allowed to order one thing. My sister always got a Green River soda, and I always got a malted milkshake. I think if I remember correctly, my mom would get a simple Coke and she would pour it over thick ice cubes with her white gloves. It's amazing the details you can remember sometimes. When we were all finished, we would go looking around the store. They had everything. We weren't allowed to touch anything, but I loved going to see the toy section and the candy shelves. I even liked to go look at all the bedding and pillows. But the best part was the elevators, of course. We would ride them up and down. We didn't get to press the buttons in there—oh no, no. There was a man working inside and he pressed the buttons for you. It was so fancy. Nowadays, the allure of shopping is all gone. You don't wear gloves for anything. You press your own buttons. If you want to buy anything, you just *click click click* and it comes to your door. It's sad in a way. What I wouldn't give to go visit Wieboldt's again just to taste a Green River soda and a malted milkshake.

—Anonymous, audio-recorded story

Marketplaces in the literal sense are places in which things are bought and sold. We've all been to a supermarket or a farmers market or maybe even a meat market (an actual one and not a bar filled with dudes on a Saturday night). What I love about a market is that it is a validator of worth. If there is a market for something, it means people are willing to pay for it. If farmers are selling fruits and vegetables at the market every weekend, we can conclude that people want fruits and vegetables. The farmers' prices can help us decipher the value of particular produce during a specific time and in a specific location. If people are willing to pay $5 for a basket of strawberries at the market, that market may set the standard price of a strawberry around town.

Markets are powerful places. The advent of the internet ramped up markets' potentials, and AI technology can expand markets well beyond a Saturday morning farmers market. Look at the power of one of the biggest markets: Amazon. Amazon can tell what people in Spain are willing to pay for a stroller and that people in Mexico are willing to pay half as much. This data drives sellers to certain countries and makes decisions for the biggest businesses.

Sometimes markets set prices way too high—much higher than the actual product's worth—to entice people to want it. Have you heard of "tulip mania"? It's a term used these days to describe that kind of economic bubble. It originated in the 1600s when the price of tulip bulbs soared. When certain bulbs were newer or harder to find on the market, their prices rose. At the height of tulip mania in 1637, the price of a bulb had ballooned to up to ten times the annual income of a skilled artisan. And we thought Erehwon was expensive! People bought them because people love to get what's hard to get, and the market happily obliges.

Modern-day markets are still setting prices and giving us a place to spend. Airbnb is another great example of a relatively new kind of marketplace. Their algorithm works hard each day to test out pricing and see what people are willing to pay, defining the prices for the next day. In fact, there are several pricing software add-ons you can use when "selling" your vacation rental homes so that potential buyers can be charged the best "market rate."

Companies like Airbnb and Uber, or the myriad of others, have "disrupted" industries like hotels, and yellow taxicab companies have cut out the middleman, providing the consumer with a sense of control.

Originally, if you wanted to visit a different city, your only option was to stay in a generally overpriced hotel in a touristy area. Airbnb provided tourists with a variety of options. Airbnb also gave entrepreneurs or homeowners the opportunity to rent out their homes and earn more income. Airbnb took the hotel market and created sort of a subset of a hospitality market that people loved even more. People jumped at the chance to try it out, travel more, and rent out their own rooms or houses. It was a win-win for buyers and a new kind of seller.

Yes, Airbnb still made a ton of money as a company. It also provided a viable extra source of income for people, and a plethora of new side hustles. Uber's pioneering model also empowered drivers with a new side hustle that did not have the same stringent requirements for earning a gold medallion to become a Yellow Cab driver. Consumers didn't have to wait until a Yellow Cab they were flagging stopped for them or wonder if they were getting ripped off as the meter ran, as prices on Uber are predetermined and set before you click the button to schedule your ride.

New businesses like Airbnb and Uber are lending themselves toward something I like to call the democratization of a market because they give power back to the people, something we desperately need in several sectors.

Mother Jones has a bit of disdain for the idea of democratization pertaining to industry, and their reporters roll their eyes at everyone doing it:

> Glossier, a cosmetics company, explains it is "giving voice through beauty" in order to "democratize an industry that has forever been top-down." Robinhood, an app that gamified trading, says its

"mission is to democratize finance for all." CoachHub, a corporate coaching company, asserts: "Our Mission: Democratize coaching." It goes on like this on About pages. Airtable wants to "democratize software creation"; Bolt is going to "democratize commerce"; PayPal is working to "democratize financial services." Elizabeth Holmes, infamously, set out to "democratize healthcare."[1]

The idea of democratization is everywhere! If everyone is doing it, I say let's jump on the bandwagon. If companies like Robinhood are making it easier for regular people to become investors, why the hell not? If Airtable wants you to be able to make your own software without having to hire a developer, yes, yes, yes! Then there's Swimply, which gives people the power to rent out their pools. Peerspace is for event rentals. Dating apps cut out the matchmakers. Everyone really is doing it and experiencing great results. You can do things as an individual that used to take entire systems and teams to facilitate.

Industries that have figured out a way to cut out the middleman and create a democratized marketplace that gives the power back to the people are harbingers of good things to come. They reinforce my idea that a democratized marketplace could help give agency to societies where people are losing power and sovereignty through technology. A democratized marketplace could help bridge the widening gap between rich and poor.

The Lotic marketplace would be a place where we could democratize ourselves. We could tell our stories and then break ourselves down into patterns and biases and decisions, using them for more awareness, to get help making decisions, and to make purchases. This cuts out several kinds of middlemen.

Right now, we still have to use a team of people to get to know ourselves better. We need support groups, self-help books, and many of us need psychologists or psychiatrists. Even with all this assistance, we still fall short! We still look away instead of asking ourselves the tough questions. We still hold onto high expectations that end up disappointing us.

What if we could democratize that self-knowledge and use technology to help us know ourselves so well that we could predict our biases long before they came into play, or understand that we're predisposed to falling for certain scams or have a pattern of quitting before we have a chance to win? What if, instead of understanding that we like red Toyota trucks like the data companies might tell us because we clicked on something once, we could understand what kinds of thoughts would make us feel like we needed a truck in the first place? Democratizing ourselves gives us back the control we've lost when it comes to knowing who we are.

In this marketplace, the storyteller earns a token for every story told, as long as it is different from the previous stories told. You can't just tell the same story of that one time you got caught stealing a pencil in second grade over and over again to gain spending power.

Earned tokens could be used in two ways:

1. **IN THE MARKETPLACE:** Brands and service providers can use the marketplace to find customers and gain loyalty, or they can choose to sell their products and services directly in the marketplace. Lotic users will be able to spend their tokens on offers. Say you fall in love with that recycled gold necklace. You can buy it right there. This makes the Lotic marketplace seem like a more conscious, caring version of Amazon. The difference is that it offers new opportunities to those who might not have had them before. Say you've never had the budget to take good care of yourself. You might be able to use your stories to connect with healthcare providers who give discounts to those who can't afford insurance. Or the Lotic tokens can be used to book massages, buy smoothies, or do other self-care activities you wouldn't have even dreamed of being able to afford.
2. **TRADE-INS:** Because the entire platform will be built with the same decentralized ledger and Blockchain technology used by cryptocurrency, I imagine a time when the tokens will be traded for actual cryptocurrency. This gives people a chance to invest in their futures or to literally make something out of nothing but their stories.

Brands or service providers opt in to both purchase data and to send curated messages to those who opt in to see them. Both sides retain some power and can decide to dissolve the brand/client relationship at any time. Already bought that gold necklace and don't want to spend money on a gold necklace ever again? Then stop sharing data with the company that sold it to you and never set eyes on their messaging again. Already tried the protein frozen yogurt, loved it, and have become a weekly customer? Stop sharing data with the frozen yogurt folks and get their messages off your screen. You have the power to decide to not be talked to at all and can simply share your data with healthcare providers who might help you through appointment care instead of selling you something.

The democratization of ourselves creates a marketplace of power where stories are king and both sides reap benefits.

To break it down even further, the Lotic marketplace works like this:

1. People record raw, in-depth stories into their devices any time they feel like it.
2. AI-generated prompts help them reveal rich information that they give freely. They have the opportunity to delete the story if it makes them feel too vulnerable. They also have a chance to save the insights for use as currency.
3. Those who tell stories receive tokens for each story.
4. Stories are encrypted using Blockchain technology for everyone's security and privacy.
5. The Wisdom Engine parses out important data points and insights based on your stories.
6. Insights gathered from the stories help us learn about ourselves. We uncover deep-seated patterns and finally heed the advice from Plato to "know oneself." We become better decision-makers and more aware of what we truly want.
7. Marketers, brands, healthcare providers, services, and other companies sign up to sell their services in the marketplace.
8. People find goods and services in the marketplace tailored to them based on their rich story data and insights.
9. Companies become 70 percent more efficient by using individualized insights, saving themselves millions in customer acquisition

fees and manufacturing costs. They reallocate a portion of those resources (profits) to provide value to the individual.
10. People use their stories for purchasing power or they trade them in for actual cryptocurrency.
11. The story economy is born, and people can opt in to an alternate economy if they so choose. We all have more options and agency.

CHAPTER 18

The Context

I don't know what happened but Google somehow connected my Gmail account with my boyfriend's Gmail account. We didn't realize it until it was too late. My boyfriend was announcing his new business to the world and so he clicked *all* of his contacts when it was time to send his very first newsletter. Unfortunately, *all* meant all of his contacts and all of mine! How did he have my contacts too? I have no idea. But he sent a newsletter announcing his new business to all my ex-boyfriends, all my coworkers, all my business contacts, and every friend I've ever had or every guy I ever flirted with and exchanged emails. It was humiliating. Thankfully, he didn't have my name anywhere in the email, so I hope everyone thought it was spam. However, one guy wrote back angrily demanding where he got his email. It was a guy I'd gone on one date with, and he lived kinda off-grid, so he was so mad and sent several fiery responses until I finally told him that I don't know what Google did, but I swear it was Google's fault. How did that happen? It's crazy that a company that knows so much about us can just accidentally share every single one of my contacts with someone else. I never ever figured out how to take those contacts away from my boyfriend, but thankfully he is now my husband so it's not such a big deal anymore.

—Anonymous, audio-recorded story

The technology currently in place can track us with the creepy accuracy of a tracker jacker in *The Hunger Games*.

How would the material the Wisdom Engine gathered be better than the terabytes of data already out there on you already? What could a spoken word story tell us that the data brokers don't have on you already?

A lot. I keep going on about how the Lotic data and insights would give richer, more intimate details about you, but what does that even mean? I'd venture to guess your audio-recorded stories might not gather the terabytes to rival those of the data brokers. However, the quality and context of the story data the Wisdom Engine collects could be exponentially more interesting and useful.

My maternal grandfather, Everett Rea, was a sergeant in the US Army Air Corps, assigned to the 803rd Engineering Brigade.

He entered the service in March of 1941. A quick search on him will tell you that he was a prisoner of war in the Philippines and returned to the US to marry his high school sweetheart and went on to have three children and ten grandchildren. You can even find his obituary and see that he retired in Florida and had a loving nature.

Data brokers know all of this.

They would never know, however, how strong he had to be to survive his three-year imprisonment during World War II, which included the infamous Bataan Death March. He was taken prisoner in 1942 when the Japanese Army forced around seventy-two thousand POWs to march from the Bataan Peninsula in the Philippines to Camp O'Donnell, a sixty-five-mile journey through tropical heat, humidity, and rain, where soldiers encountered starvation, heatstroke, broken bones, sickness, and extreme thirst. If anyone fell out of line, fainted from hunger, or even stopped to take a breath, they were instantly shot by Japanese soldiers. My grandfather made it through. Around eighteen thousand soldiers did not.

He was held prisoner after that and forced to work in mines in Japan until September of 1945 when the war ended. His mental strength kept him hoping and trying to survive. He arrived home severely malnourished and with broken bones.

He died in 2010 but imagine what the stories he'd tell about this time could reveal about him. They would explain his appreciation and gratitude for every moment. They would record his fears, his relief, and the fortitude it took for him to survive. They would attempt to describe how he moved through life, why he made certain decisions, and what he valued most.

Those stories, if he'd recorded them alone in a private interview with expert, inquisitive prompts, would reveal his true nature, his strength, and would explain how he came up with his tagline: "I never let the past define me. I always try to look forward."

That story would be the definition of powerful. One of my biggest regrets is not having recorded it or having asked the prompts myself. While I can still listen to my mother share her memories, there's nothing like the insights you can gain from a raw story shared out loud without any editing.

Those are the types of stories that a data broker could never capture by monitoring my grandfather's internet activities or even his location.

We don't actually know what data brokers or Big Tech companies know, but we do know that their knowledge is pieced together from information gleaned from monitoring us (that is, stole when we weren't looking). Imagine they're trying to re-create your image with those Lego Duplos made for little kids. While they might be able to capture an idea of what your face looks like in primary colors, it would still end up kid-like, clunky, and with obvious Lego edges.

Yes, they can look at your search history, where you're going, where you've been, what content you love, what groceries you buy, what kind of financial transactions you make, what kinds of politicians you support, what friends you come close to, and what you value the most. And they can do a lot with that information. However, it's a clunky, Lego Duplo story, and it can never truly be a complete story because what the data brokers know about you only explains the story of the "what." It might show *what* you buy, *what* you like to do, and *what* places you visit. This type of data doesn't really get to the bottom of *who* you really are or *why* you do those things.

Sure, it gives a good picture of who you *might* be, and the picture of who you might be can look pretty sharp. The TikTok algorithm is good at feeding you exactly what content you want because they have a picture of you that is fairly accurate. The data they collect about you—your likes, comments, follows, and how long you spend on a particular video—determines what videos appear on your "For You" page. They get it so right that it can be addictive. Still, that algorithm doesn't provide much context—only the *what*.

They don't really know *who* you are. They don't really know *who* anyone is.

You wouldn't be able to stand up at a funeral and talk about the dead person's likes, comments, and follows. While TikTok might seem like they know you well, they're only learning a single story about you and seeing you as one of billions of customers. It's a story of your clicks and not your humanity.

A full story, shared willingly, that is unable to be edited, is raw and real. It uncovers the *why* and not just the *what*. It might tell *why* you always click on content about cottage cheese recipes and people cleaning their rugs. However, it will never know that you chose this content because you have pre-diabetes and your parents are ex-hoarders so you like to see uncluttered rugs. Your real motivations make for a much more interesting story.

Data farms everywhere might have a picture of your *what*: You buy an awful lot of ice cream. They don't know, however, that your *why* is that you hand out ice creams to all the neighborhood kids who come to visit for the weekly ice cream social. They don't know that you started doing that because one family's parents died of a fentanyl overdose and their three kids were completely lost, so you started reeling in the community with food to help those kids feel less alone.

The data farms might know that you buy a lot of sneakers, but they don't know it's because you walk every single day to avoid having a heart attack at forty-eight like your father had.

The data farms know that you drive three hundred miles to see your daughter in college quite often, but they don't know you do that because she called you once drunk after a party and, in a confession she doesn't even remember, told you that she was really scared of everything.

Knowing those details can tell a lot about your heart, your priorities, your values, and your needs. They explain so much more than your grocery list ever could.

Tech companies don't know you as well as you know yourself.

In a sense, they don't know jack.

The data that I think will truly change the world will not be collected by scanning your location or by counting clicks. It will be collected voluntarily via raw, spoken-word stories, so that it gathers a real picture of who you are.

The Wisdom Engine uses the ARI we developed to provide you with friendly, curious prompts that give you the opportunity to share what's on your mind. It's free association, stream-of-consciousness thinking. It's raw and unedited and it's extremely valuable information. You've seen some of the stories I've collected at the start of every chapter here. Each of them gives you insight into who the people are, what's important to them, what's at their heart. These are real, contextual pieces of information that will never be collected by tech companies unless you decide to share them yourselves.

A raw story often uncovers what I call your "lived truth." Let's say we both go to a concert and sit next to each other. We see the Eagles. I get home and tell my wife that the concert was great and go to sleep. You come home with your heart full because, the last time you went to see the Eagles, you were with your now-deceased mom. The songs Don Henley sang felt so much more meaningful to you this time around. You stay up writing poems about the experience and feel it was otherworldly, like your mom was speaking to you through song. We both can have the same exact experience, but we walk away with two entirely different stories. Our whole lives are compounded to create our belief system and give us our experiences. It's the stories that shape our everyday life. They matter so much more than we give them credit for.

A story is where the context lives. That context is how we capture the lived truth of a person's experience. You can tell me all day long that something is good for me, but if I've been experiencing that same thing all my life as wrong or uncomfortable, I'm not going to believe you. The data says you should walk every day after dinner, but if you sprained your ankle last time you walked and got lost in the woods the time before that, you're not going to think the data is correct. Our unique stories make up our personal truth, and that truth is different for everyone.

Say you grew up in a hunting family who safely stored guns in the house and used them often. You cannot imagine life without a gun. Your "lived truth" tells you that guns are a necessity. Those with a different lived truth have little experience with guns and therefore only see them in the news as murder weapons. They're scared to send their kids to school and want to ban all guns. Who is wrong? Based on the lived experiences, neither one is wrong.

Different truths can certainly divide us, but I hope that learning someone's raw story can actually build more bridges than burn them.

If we really dive deeper into the stories of both the person from the hunting family and the person who has never held a gun or never will, we might find that these individuals are more similar than we might expect. They both have a deep-seated urge to help and protect their families. They want the best for their kids. They want to feel calm and peaceful as they lie in bed each night. The issue that divides them is that they have a different idea of what brings them peace.

What if we could focus on the connective bits first when it comes to legislating instead of simply fighting about the issues? What if a politician could use that contextual data and say, "Okay, I've collected the millions of stories you've sent me and I've used AI to see that over 90 percent of you, regardless of political affiliation, feel scared for your children's health and safety"?

Immediately, that politician has our attention on both sides. Then she could say, "Let's think of more ways to make our children safer." Then we're in it together.

We could look at the data regarding guns (both historical and from more of the millions of stories collected) to determine whether or not guns are safe. Then, we determine what to do together as a society after uncovering everyone's lived truth. We'll never make everyone happy, but we can make everyone feel heard. I'll take feeling heard over the system we have now. The reconciliation of two seemingly disparate perspectives only seems possible with real data collected from our truths.

Stories can also help you build bridges within yourself. What if you shared a story every day for a month and used technology to help you uncover patterns within those stories? You might be able to see that you talk an awful lot about not liking your colleague at work. You might uncover that certain preconceived ideas you have about your colleague were based on biases you didn't realize you had. Sharing one story a week could get you to notice misconceptions you've been holding onto and help you to let go of them.

CHAPTER 19

The Insights

So, I learned something big today thanks to my therapist. I'm actually embarrassed to say this because it's so obvious that I can't believe I didn't see it. I was so in love with this coworker that I met over Zoom meetings. We sent each other texts all day and night, and I could not wait to meet him, but he never wanted to meet up. We live in the same city, so come on. We kept having these huge disagreements because he would make a plan and then always cancel. So unavailable, right? Well, I basically told my husband I wanted a divorce because of this guy. Not like I thought I would end up with the coworker, but I just knew the feelings I had for him were so much better than any feelings I had for my husband, and it was a sign that I needed to finally make the big decision to leave my husband. But we were living together. We were parenting together. We were planning meals together. Here I am accusing the coworker of being unavailable. *I'm* unavailable! Me! I cannot believe I have so much work to do on myself. It's insane. I am sad about it, but I am happy that I now have a to-do list. I will be totally emotionally available to the next person who steps up. As long as my husband has moved out by then.

—Anonymous, audio-recorded story

My main goal with this work is to create valuable insights from our shared stories that uncover biases, patterns, and expectations. I must warn you: At first, it might not feel good. People don't love learning they're wrong or that the beliefs they're holding have been harming them or holding them back. I anticipate a few defensive users, at first. "How dare you say I am wrong in my thinking!" What would the world be like if we could get rid of the filters we unknowingly place on every one of our points of view?

Confronting our biases is difficult, and up until now, there hasn't been a single magical way to clean off the glass smudges obscuring our view. Under certain circumstances, that's okay. I have a bias against standing shoulder-to-shoulder next to a hungry lion, and I won't do it. That's probably smart.

Let's explore a way that the Wisdom Engine might be able to help us uncover bias. Suppose you keep going over your monthly spending budget every month. You've listed out everything you're spending money on and cannot see any way to cut back on anything. You feel like everything you're buying is absolutely necessary. When a friend looks at your budget and sees you're spending $4,000 on rent and $500 on nail salon visits each month, she might tell you that you can cut back on those expenses. You can't fathom that suggestion, as you value a huge house and always having pristinely painted nails.

Sakichi Toyoda, the founder of Toyota, created the 5 Whys to improve business and make better decisions. The technique is a way to reveal your biases or simply understand yourself better. Toyoda said that you must go deeper into a story to find your reality. That deep dive is akin to what the Wisdom Engine accomplishes. When pondering a problem, according to Toyoda, you must ask yourself "Why?" five times.

Let's see if it works in this situation: Why do I value a huge house and polished nails? Because these things make me look good. WHY? Because they make me look richer than my friends. WHY? Because I must look richer than my friends. WHY? Because I want to seem superior to them. WHY? Because it makes me feel better about myself. WHY? Because I actually don't feel so good about myself. WHY? Because I feel not as smart as other people. WHY? Because deep inside I think I'm a loser. WHY? Because that's what my dad always told me. WHY? Because he probably felt like that and wanted to bring me down with him.

Aha!

If you did this with every single problem, you might be able to change your current reality for the better. I do wonder if sometimes people have no interest in changing. I mean, we've known about Toyoda's business strategy for decades, and it's not very often we use it to help ourselves detect any biases. If we took the time to do so, the wins could be huge!

If you could put the thought that "you're a loser" or the idea that you need a big house and expensive nails to prove your worth to the side, what could happen? Say you put that thought away in this situation. You find a much smaller place to rent for $2,000 per month and begin to paint your own nails. You feel a rush of relief at being able to live within your means. You might feel more powerful, begin learning to invest your savings, gain more confidence, and be able to quiet the thought that deep down you're a loser. It might take some hard convincing and an open mind for you to become more consistently accepting of your new lifestyle, especially if your beliefs equating money and image with worth were deeply ingrained. It sometimes takes an unlearning to get where we want to go. In the end insight could help reacquaint you with your true self, and lead to unexpected changes in your life.

This example uncovered an inner narrative and a bias. The inner narrative was that "I am not good enough" and the bias was that "successful people have painted nails and nice houses." So much can change by simply becoming aware of these details.

We're training the Wisdom Engine to go even further and to uncover expectations and patterns, two other areas that can help with our decision-making.

I have heard that expectations are planned resentments, and I believe that's true. How often are we disappointed by our expectations? There used to be a popular BuzzFeed video series that explored the difference between expectations and reality. Your expectation might be that your first job will be fun and give you financial freedom. In reality, it's hard, boring, and hardly pays anything after taxes. Having even a normal expectation like excitement for a first job can sometimes lead to huge disappointment.

We are often let down by Pinterest cake recipes that end up looking like soggy sandcastles or by larger concepts like adulthood—it's way harder than I thought it would be!

I happen to be quite curious about various religions, and I've met a lot of people who had expectations about their own religions. They often thought, "If I act in a devout way and follow all the rules within this religion, then I'll be taken care of and nothing bad will happen." So they do that and have this idea of how the world should work, but it fails them. They do all the required things, and the world returns something they didn't expect, and then whoops . . . they shatter. Understanding the truth about how the world works is essential. You need to see the world clearly. Unrequited expectations are at the root of so much pain.

I'm not saying to never set an expectation. Go ahead and shoot for the stars, expecting to win! Idealism is what's making me pursue this business and write this book.

I'm simply suggesting that expectations be made after some self-reflection and uncovering of biases that are guiding us to see things unclearly.

What if we could use our stories and AI technology for self-reflection and to better understand our biases and reality, so our expectations were set in that context? There would be less pain and disappointment in the world if our psychological burdens were eased. High expectations and disappointment can be a pattern we continue over time, like other patterns that can sometimes run our lives.

Some spiritual teachers say that it's your job on Earth to learn lessons, and you'll continue to be presented with the same lessons until you finally learn what you're supposed to. I don't know if that's true, but I do know it's common to fall into the same kinds of patterns without realizing it. We fall for a similar type of person even after we swore we would never date someone just like our mother again. We buy things and feel remorse over and over again. We continue to find ourselves in jobs that require more hours of work than there are in a day.

By analyzing your stories over time, the Wisdom Engine can give you a list of your potential repetitive patterns. Do you keep helping your friend out even when you don't want to? Does your dad keep asking you for money and you keep saying yes and resenting him for it?

Imagine being aware of these patterns and having the ability to say, "Well, I would like to say *yes* to a date with you, but I see that you are similar to the other people I have dated in my patterned past dates, and I would like to make a different choice this time around."

What kinds of things could we change in our lives if we had a technology that could help us with these big realizations? Who could we be? It's exciting to imagine, especially if it's presented to you in a dashboard for you to take in at your own leisure. What if you're paid to do it, or if you receive tokens for learning about yourself? The scenario is a win-win-win.

CHAPTER 20

The Ledger

I got divorced in 1992. So, it's been a while. I am so happy now. I have been happy every day since. It's wonderful. I was married for twenty years. That's a long time. And during that time, I took everything on myself. If we had a fight, I would look at where my responsibility was in it. If there was a problem with finances, I would take a look at how I could help. It was work. Hard work. And you know what? That's okay if both people are working the same amount. If two people are in it together and they can come to an agreement about everyone's roles, great. Great! But after twenty whole years of my short life, I realized that I was the only one doing all that hard work! One day my friend told me, she said, "Sure, you are working on everything, Rachel. But you know what? It's okay to also take a look at the part that is not your responsibility too. It's okay to call that out." I had never considered that before. I always thought everything was my fault. I don't know where I got that stupid idea. Well, I spent the next year observing everything after that. And I realized that I was doing it all. I didn't really have a partner in the slightest. So, I got out of there and I would never get married again even if someone paid me a million dollars. Not like they would. I'm seventy-five now—ha! And boy am I so happy that my world is so peaceful.

—Anonymous, audio-recorded story

My biggest priority in creating Lotic is to give humans back our power. I want us to remain in control, and I want our tech to be used as a tool. This requires a unique way of storing the stories and converting them into currency to be used in the marketplace.

This storage system might be of more importance than the insights we get. What is the point of understanding your biases and becoming a better human if everyone can see those biases and vulnerabilities and use them against you? We must ensure that no data broker can sneak in to grab this data or trick you into signing away your rights to it.

This is where it gets a bit futuristic. I promise I will not try to sell you on cryptocurrency but brace yourself because I am about to talk about Blockchain technology.

The underlying technology of the Blockchain is what's called a "decentralized distributed ledger"—a database hosted by a network of computers instead of a single server, making it impossible to hack. Your stories will be safer than if they were in a bank. Banks use a centralized system to store your banking data, which can be risky.

With the functionality of the bank or a service such as Venmo, two people agree to exchange their money. In this case, the bank acts not only as the third-party service provider that supports them with money transferring but also as the keeper of the records, storing proof of each transaction. Therefore, the bank here can be regarded as a centralized point of control.

The problem with this is that, if anything bad happens to the bank or the third party, all their users will be affected. For example, if the bank's server goes down at the time a transaction occurs, it might not go through successfully, funds might be lost, and you might lose the record of that transaction. If the technological glitch is severe, you could lose all your banking records.

Since all your financial information is in one place, it takes only one hacker to get into it, steal information, and there goes your retirement savings! This fear is one reason cryptocurrency offers great promise—as a way to safeguard our money by holding the information about it in several places, not just one. It's Blockchain's mission to do the same, dividing ownership of information among the many users instead of on just one server.

The name "Blockchain" comes from the way data is stored in this decentralized system. Right now, on your computer, I'm assuming you have a bunch of folders you can open with files inside them. Let's say you click on the folder called "SUMMER VACATION," which opens so you can find a bunch of photos of you and your family frolicking at the beach. If you want to send a photo to your grandparents, you can click on the photo and send it along. Or, if you want to delete an incriminating one, you can find it and put it in the trash. It's fairly straightforward.

Blockchain technology would never allow such easy sharing or deleting of files. Every bit of data is stored sequentially, and it's attached to the piece of data before it. Imagine all those summer photos linked together in a chain. You can't send your grandparents just one photo because it's linked to all the others. Because it's all connected, it's also stamped with the exact moment it became linked, so you know when it arrived in the chain and that it hasn't been touched or moved. This sequential coding provides a high level of security, and it's useful when proving something is an original. It can't be moved from the chain, so it's easy to see exactly when it was created.

It is impossible to change this data, fake it, or steal it, making it the safest way to store it and the best way to retain your power over it. Every time a story is told, a contract is automatically written. It passes through the decentralized ledger, hits the Blockchain, and a token is issued that you can use for spending. Plus, in our system, all personal details are subtracted from the story before it is stored. So, even if some bad actor figured out how to get through the impenetrable, they would only be able to find a story but with no name or personal details attached to it. If it sounds like a hundred steps of security, it is.

This might seem a little extreme when we could choose the easy route and simply save your stories on a server. It is. We are, however, asking people to get comfortable being uncomfortable as they share their most private personal stories, so we must provide assurance that these details are safe.

To me, data sovereignty is almost the entire point of doing this. I want us in control. So, any possibility of a data breach is out of the question.

I want to ensure that all the data pulled from those stories will remain in your possession. Stories or data gathered from the stories cannot get sold to anyone else. Companies cannot steal bits and pieces of your information to create some unknown profile of you that a data company holds secretly and sells (like they do now).

As I've stated before, if my version of this idea doesn't work out but someone else runs with this idea, I will be a happy man. However, this step is a step that cannot be skipped. The data must stay at the individual level and in *the individual's* hands. Otherwise, it will just get sold and sold again just like it is now. Nothing changes.

CHAPTER 21

The Future

My son and I volunteered to fill up Easter baskets last year. It was with a great organization, and they had all these donated items in different areas. It was our job to go around and pick up different things, arrange them nicely in a basket, and then go wrap them with cellophane and ribbon. It was fun. We filled so many baskets and after a while the room was packed with them. My son looked around the room in awe and then he got a confused look on his face, and he goes: Dad, why are we making Easter baskets for kids? Why doesn't the Easter Bunny just stop at their house too? I was like, ummmmm . . . I ended up making up some story about how these were for teachers who wanted to give them out at school. I don't know if he bought it, but he had a really good point. Why can't the Easter bunny go to everyone's house? Why is life so unfair sometimes? It bummed me out because my son is so sweet now (he's six), but one day he's going to know how unfair life is and that just breaks my heart.

—Anonymous, audio-recorded story

The entire future of the story economy hinges on the idea that people will use it. Brands need to buy in because the people are there, and people need to buy in because the brands are there to hand out those tokens. So, a bulk of my work now is to create a product that everyone loves, and then let people love it.

So far, we have Lotic running a basic beta version as a progressive web app, and we're working on a wearable piece of tech that people can use if they're not interested in bringing out their phone for yet another use. Storytellers can already earn tokens and learn about themselves, but we're currently working on exactly how to implement a true exchange of goods and cryptocurrency for story tokens.

My biggest hurdle now is to convince people to try it. I can write this book. I can speak on podcasts. I can make ads and pay influencers to boast about it. I can cross my fingers that it all works. But, if it doesn't . . . the entire idea fails.

If I only capture the attention of a million people, the idea falls short. The utopia on the other side of this technology requires mass adoption. Mass adoption of a technology that allows for a new currency—a currency that we innately carry within us—would transform the way the world assesses worth and sees our value. We'd experience an equality that's been long forgotten. Sure, we're humans and we'd find ways to divide ourselves, of course, but what if we all walked around feeling valuable and recognizing that others are valuable as well? What would that do to society? I want us all to see the value in our pasts and how we talk about them. I want us all on equal footing. I want us to have access to everything we need and power over our own data. If I suggest this shift as an option, how could we all not want in?

Lotic, or any other company that can offer a story economy, entices customers with 1) financial or economic enrichment, 2) technological security, 3) control over our personal data, and 4) a richer sense of self. I personally would run to be an early adopter of this technology if I had heard those selling points. (I also had an early Oura ring and have recorded millions of other health data points on my body, so perhaps I am way too biased to tell whether people will flock to experience a story economy.) However, I have been studying human/computer interactions for almost two decades across a bunch of different types of use cases, data types, and digital/technological systems. Through extensive research, my colleagues and I have continually found that people are open to different ways of using their technology as long as they can see the value in the fresh model. In other words, humans are receptive to different ways of engaging their technology and tend to show preference toward usage types that offer the most personal value (monetary, emotional, dopamine, etc.) as well as those that reduce friction within the person's life. Value and convenience lead the way every time. A story economy delivers on value and convenience.

I saw this with a current user of Lotic. She answered an anonymous survey, but we call her Liz. She's been using one of the first rollouts of Lotic for months and has made it her daily journal. We began by paying her for stories with tokens that can be traded for Amazon gift cards. It's not the marketplace I envisioned quite yet, but it offers a similar incentive. Liz says it's the Amazon appeal that got her interested in doing the daily check-ins. However, she noted some surprising awarenesses. For example, talking out loud about some issues she was having at work helped her "realize that I was getting short and annoyed when people asked me to do something they can clearly do for themselves." She said that through the platform, she was able to make a goal to simply "be helpful even if it's a silly ask." She says that "a couple relationships at work have taken a turn for the better. It's like I can appreciate the person more. See them more as a human."

She didn't say the system is perfect. She commented that we overuse the word "values" and "it's annoying." However, unprompted, she noted the work of the ARI and how she felt like a friend had truly seen her. "I answered a question and it processed it and the statement that led to the next question was so simply put and so meaningful to me, I've been using it at work all the time: 'You took a proactive approach and turned a potential conflict into an opportunity for collaboration.'"

When I read the answers from these anonymous surveys, I feel gratitude and a little bit of shock that our AI said something so meaningful to someone that I've never met. It's actually happening!

Awareness has always been part of the plan. Understanding our biases has always been a major part of the reason to do all this in the first place. Still, I am shocked that it's what some people focus on the most. Liz, for example, mentioned Amazon only once over fifteen questions. She is in it for the self-knowledge. How cool is that?

Sure, the setup isn't quite complete so the survey is skewed, but I was imagining people would be most interested in using the marketplace as a way to be treated like a millionaire. Let me explain.

Imagine two people standing next to each other. One makes $10 million per year. One makes $10,000 per year.

The whole global economic system has been articulated to serve the person who makes $10 million. We're living in a $10 million world without really realizing it. The world caters to them, makes things easy for them. Those with that much money can get what they want, how they want it. They snap their fingers and get healthy meals prepared. If they feel like they're gaining weight, they can get a trainer or have a gym created in their home. If they need to remodel their kitchen, no trips to Home Depot to look at faucets are necessary. The wealthy homeowner has someone bring in samples for them to choose. They don't really need to fret over the price. Everything can be totally personalized for them. This reality doesn't just apply to faucets. It applies to healthcare, schools for their kids, and transportation. Those who can afford the best of the best get it.

What about the person who makes $10,000 per year? They can't get any of that care. They buy what is available and within their budget. They find the cheapest thing and go with it. They might buy their rice and beans from Dollar General. If they have time to work out, it's in the park instead of with a trainer. They don't remodel their bathrooms. This isn't necessarily bad, but it isn't fair. The thing is, they don't have $0. They have $10,000. They're usually ignored as a market even though there are many more people with $10,000 than with $10 million. The $10,000 people don't get the attention or market spend the richer people get because most brands aren't sure how to market to someone with such limited purchasing power. Yet, they're missing out because the amount of limited spenders is huge.

What I'm forecasting is that a story economy will open new markets, an opportunity I didn't even imagine when I sketched the marketplace in the beginning. It will give lower-income people more buying power, creating a fresh demographic ready to meet new brands. The marketplace will also thrust those people into a new category in which they will enjoy a more personalized shopping experience, previously offered only to those in higher income brackets.

Using that rich story data may make buying and selling easier for the market, the marketers, and the buyers. That $10,000 person could get just as much personalization as the $10 million person—a digital, mini version of your interior designer presenting you with fabric swatches before you redo your living room.

With your valuable stories recorded and analyzed, the algorithm is then able to personalize products and options just for you. So now, suddenly, there is a list of goods and services that are targeted toward the $10,000 population. As a $10,000 person, you can get much closer to that personalized experience the $10 million person has. You get to see a broader variety of options and possibilities tailored for you. Instead of being forced to flip through ads for the new New Balance sneaker you can't afford, you can now see the $10 version of the sneaker that New Balance has produced for people in your market. Or you might see a broad variety of options just like that New Balance that have been specifically manufactured for someone with your budget. Brands are more willing to make something so personalized because they can use the story insights to calculate that it will sell.

In the same way that the $10 million person is presented with options by their contractor or house manager, the $10,000 person might be presented with plenty of viable options as well, growing their purchasing power.

This also applies to healthy food. Organic veggies and $15 ketchup are marketed to those who can fill their cart at Whole Foods. That fancy ketchup brand might jump at the opportunity to create a less expensive version for those who still want to be healthy but may not have $15 to spend on one bottle. They can get a smaller bottle or perhaps a version that uses organic tomatoes but not necessarily be packed in a glass bottle. Selling a higher volume of these to the lower-spend market will make up for the lower-priced item.

What about healthcare? My friend owns a pet clinic in Panama and opens it up once a month for those who want to pay through donation. They're extremely busy that day, but they end up making a lot of money and helping a lot of pets! What if a dentist could see in the marketplace that twenty $10,000 people were willing to come in on a specific day if she took donations or offered cleanings at a lower price? Having that information would allow for a day full of bookings and might even leave the dentist making more that day, creating yet another win-win.

I'm imagining a world in which everyone has more opportunities to take care of themselves. In these scenarios, I'm still assuming you're paying for all these goods and services with actual dollars. What happens when you are able to spend stories?

Spending story tokens may allow you to afford anything in the marketplace. If you want to live richly, maybe you save up your story tokens and finally purchase the Eames chair you've had your eye on. Perhaps you don't want something so extravagant, but you want to take care of yourself. You use your tokens to buy massages, protein powders, and workout classes. These are priorities you couldn't have had before as a $10,000 person.

This puts us all on an even playing field. We all have unlimited stories to tell. We all would love to buy things with a currency that doesn't come out of our bank accounts. We all would love more personalized services that make us feel like we have the purchasing power of someone with $10 million. Every single brand out there wants an increase in efficiency and the specific data that ensures them that what they're making will sell.

This process differs from anything that currently exists in Big Tech because our system values the customer for who they really are. Currently, we are asked to conform to other people's standards (either on social media or IRL). We spend a lot of time trying to fit in, and we are often viewed as a monolith—just a bunch of people willing to click and buy.

Lotic values your story, which is a representation of your uniqueness. You will be the only person with your data set, and brands will be forced to celebrate and cater to your uniqueness. With this plan, you are no longer just a walking wallet willing to spend, spend, spend. You are a series of stories that make you absolutely unique.

Not only will you be able to buy things with stories, you're able to continuously convert those stories into more self-awareness. You might recognize that you've had blinders on when looking for friends and suddenly you're making more meaningful friendships. You might realize your voice is valuable and speak up in meetings even more. You might learn about yourself in ways you never thought possible, projecting yourself into the life, relationship, and career of your dreams. You might learn more self-confidence, self-love, self-gratitude. Who knows what might come out of an entire population exploring epiphanies every day?

I'm writing this book at a time when America seems more divided than ever. The gap between wealthy and poor is as wide as it's ever been. Many of us can't stop to think about any of it because we're too busy trying to make ends meet. We're too lonely. We're too broke. And we're too lost.

Elon Musk predicts that AI will take all our jobs. He has suggested a different alternative economy, that of universal basic income (UBI).

Alternate economies are coming. While it seems like receiving money from the government each month might be nice, that still creates a "loser." We'd have to take money from somewhere to give it to every citizen. As I've mentioned, this can lead to resentments or shortcomings in other areas. Plus, in a UBI economy, we as humans might struggle with a lack of autonomy, losing out on choices with such limited spending power. You should know by now that agency is the name of my game.

In a story economy, everyone contributes something, and everyone receives something. Plus, we are valuing stories, a very human currency we each hold dear. There is potential to use our past and present to take back our power and retain the autonomy that is currently flailing to stay alive.

At mass adoption of the story economy, I imagine a world where we can finally let out a sigh of relief knowing that we have currency in our wallets and we can take care of ourselves. I imagine a place where we don't have to hustle to feel valued—where everyone feels praised and rewarded for their uniqueness, where we no longer fear corporations because we know they're out there working hard to get us exactly what we want and serve us well.

I imagine a world where we feel like we have a friend in our pocket helping us make decisions; where we can see our lives clearly and have help managing our emotions and understanding our patterns so we can break free from the harmful ones; where Bernie Sanders can stop urging us to close the wealth gap because we all have an opportunity for equal spending power.

I imagine a world where we can realize how much we have in common through data that shows we all carry the same hopes in our hearts despite feeling so isolated sometimes.

Where we feel safe, healthy, and optimistic about our opportunities.

Where we feel taken care of.

Where we feel like we matter.

That's the world I'd like to live in.

CHAPTER 22

The Tasks

It's normal for people to give up. They give up all the time. I want to give up too. Life seems hard. Look at the news. Nobody is having kids. The weather patterns are scary. Politics is getting really weird. Everyone's fighting. But I gotta put my big boy pants on and my fake-it-till-you-make-it smile because I got a big family. I got a momma who doesn't need to know I want to give up inside. I got four small nieces who look at me with so much hope in their eyes. I got a helluva lot of dogs counting on me and a bunch of friends who look to me to make them laugh. I gotta keep going no matter what. Maybe they need me. Maybe I need them. I don't know. All I know is I gotta not listen to the news and I gotta keep going and keep trying and focus on the positive. It's all I can do right now.

—Anonymous audio-recorded story

I've just presented you with a society-shifting plan, nine years in the making. I believe it can solve some of our deepest issues. We can heal our loneliness by knowing ourselves better, clearing our biases, and learning to open ourselves to new things. We can beef up our spending power when we use our never-ending stories as currency. To feel less lost and take back control, we can make sure we understand what data is being collected about us and have the ultimate say over how companies can use that data.

Even though I've been working on this idea for so long, I am still early to market. I can tell by the reactions of people, when I tell them about the story economy, that many don't totally understand the need for alternative economies just yet. We still have our jobs and our comfortable houses—we can't imagine what it might be like when there's no longer a need for our work or the careers we've worked so hard to build. I hope that, by the time you read this book, the idea has been widely adopted. I hope you have heard of Lotic by then. However, this idea is bigger than Lotic. I almost didn't want to tell you the name of the company I've founded; I'd hate for you to read this book, do a quick search for the company, and find out it died on the vine because it was too far out ahead of where the world is, which is quite possible. I want this idea to succeed, Lotic or not, because I believe it's the only way forward. It's the only way we can truly take our power back. Plus, it will take more than Lotic to work out. It will take you deciding to give it a try and brands deciding your stories have value. It will take all of us to make it happen, and I hope this book can inspire us all to put in that work.

If I haven't convinced you yet, what about the idea that simply using a new technology like a story-based marketplace might be the way we can actually change public policy? Since we now understand that lobbying for good within the tech/privacy sector is quite an uphill battle, it might be comforting to know that policy often follows innovation. When something fundamentally changes the way that humans can, and want to, live their lives, lawmakers often regulate it after. The policy lags behind, and for good reason; otherwise, innovation would be stifled on a regular basis. For example, when Apple released the first iPhone, there was very little policy around the effects and impacts that might be experienced now that people were walking around with supercomputers in their pockets. Policy followed behind to address things like privacy and security. Apple baked a lot of those things into their systems from the start, and, thus, it was more straightforward to develop the policy environment around their ecosystem.

Google, with the Android Operating System, did this a bit differently. They did not bake things in from the beginning, and they struggled with how to proactively implement policies and practices.

Lotic (or a company or set of companies like it) will be setting the stage for new market forces and behaviors to emerge. By being deliberate and thoughtful with our approach toward privacy, security, and data collection, we are attempting to put in place optimal conditions for establishing, starting, maturing, and operating a new type of marketplace—one centered on the conceit that the individual has agency and thus skin in the game. These forces don't exist today. These forces could set a higher standard and potentially force policy to follow. Why allow Google to secretly monitor your emails and have no transparency surrounding their data-gathering practices when a marketplace exists whose goal is to give you total control of your data and keep it privately concealed in a Blockchain?

What if these ideas forced the other tech giants to keep up and change their own methods? What if having a choice to use an application or tech product that ensured your data was absolutely safe meant that users of that product would notice a difference and demand that others follow? What if seeing those safety options and data privacy practices would prove to legislators that Big Tech's claims about how regulations would be impractical or impossible are wrong?

It could happen. It might happen. It's obviously my hope that it *will* happen.

But it might take a while. Some new technologies catch on like wildfire. Others, much more slowly. While I hope millions will read this book, love the idea, and tell everyone about it, I understand I might have to wait.

So, what can we do in the meantime? What *else* can we do to shift some of the issues and take our power back? We've lost a lot of agency, but not all of it. As I wrote earlier, Big Tech has greased the pockets of legislators, making federal regulations tough. However, there's still hope at the local level.

THE LOCAL LEVEL

We might have our hands tied at the federal level, but I have hope for change at the state or city level.

Most governing bodies, including everything from a school board to a state legislature, are familiar with the issues, and many are open to new ideas. So, if you don't want to fight the uphill battle of debating Zuckerberg in front of Congress, you can convince your library board to outlaw phones, as several have already.

Beginning with Connecticut in 2023, twelve states have addressed teen social media access. Connecticut is now requiring platforms to gain parental consent for users under sixteen. Louisiana, Texas, Maryland, and Utah followed in 2024 with varying approaches—from Utah's "social media curfew" to Maryland's "Kids Code" that bans personalized content for users under sixteen.[1]

In terms of privacy, California was the first state to mandate data collection restrictions with its California Consumer Privacy Act (CCPA). It regulates the collection, use, and sale of personal data by businesses, and it mandates specific rights for California residents to help protect their personal privacy. The law primarily focuses on transparency, consumer control, and accountability with respect to how businesses handle consumer data.

Consumers now have the right to know what personal information is being collected about them. This includes the categories of data and specific pieces of personal data collected. Businesses must also disclose the purposes for which the information is being used. Consumers also have the right to request that businesses delete the personal data they have collected about them. Businesses are required to honor these requests unless the data is needed for a legal reason such as complying with a contract, legal obligations, or for security purposes.

Consumers can opt out of the sale of their personal data. If a business sells personal information to third parties, it must provide a "Do Not Sell My Personal Information" link on its website, enabling consumers to easily exercise this right. Businesses are not allowed to discriminate against consumers who decide to click that link or ask for their data to be deleted.

This is such an inspirational start. If California can do this, why can't all the states? Some friends who live in California report a deluge of opt-out notices that came in the mail after this bill passed. Businesses are complying, and changes can happen.

California's laws related to AI, however, have not been as successful.

The Artificial Intelligence Accountability Act was introduced by California Assembly member Reggie Jones-Sawyer, in an attempt to regulate the use of AI in the state, particularly in areas such as facial recognition, hiring, and other critical decision-making processes.

The bill would have required companies using AI in high-stakes areas (like employment or law enforcement) to implement transparency and accountability measures. This would involve disclosures about how AI was being used, what data it was based on, and whether there were mechanisms in place to mitigate bias and protect civil rights.

Unfortunately, at the time of writing, this bill has yet to pass due to . . . you guessed it . . . opposition from several tech giants headquartered in the state.

Still, I find it heartening that people are starting to advocate locally. The writing of these bills shows that states are working to protect us. They are open to creative solutions and could be key to making sure our human rights are considered.

So if it's overwhelming to start big, start small. Start with your city council by asking that all city businesses comply with a labeling initiative. Work to ban cell phone use in local elementary schools and go bigger from there. Work can be done, and changes can be made.

Letter to Your State Senator or Congressperson

Here's a helpful letter you can write to state officials, urging them to enact regulations that protect your privacy and your rights. I've given an overview of the ask, but make it your own. Tell them about the specific issue you think is most important right now.

[Your Name]
[Your Address]
[City, State, ZIP]
[Email Address]
[Phone Number]
[Date]

The Honorable [Full Name]
[Title, e.g., State Senator / State Representative]
[Office Address]
[City, State, ZIP]

Dear Senator/Representative [Last Name],

I am writing as a concerned constituent regarding the growing influence of tech companies on our daily lives, particularly in the areas of privacy, data collection, algorithmic manipulation, and AI. As we see with rising concerns over youth mental health, privacy violations, and the lack of transparency within AI, it's clear that the current state of tech regulation is insufficient to protect the people of [State].

 I urge you to take action at the state level by championing stronger data privacy protections, transparency in data collection and AI, and regulation of addictive algorithms. States like California have led the way in privacy with the California Consumer Privacy Act (CCPA), and I believe [State] can follow suit by enacting similar protections, ensuring that individuals, particularly children, are safeguarded from the harms of unchecked tech practices.

 You can also help to initiate state-level research into the effects of social media and digital platforms on youth mental health, much like the proposed Children and Media Research Advancement Act. This research is vital for understanding how best to legislate for a safer, more transparent digital landscape.

Tech companies have long used lobbying power to prevent meaningful change at the federal level, but states have the opportunity to lead the charge on this issue. I ask for your leadership in ensuring that [State] protects its residents' privacy, promotes transparency in tech practices, and holds companies accountable for their impact on public health and society.

Thank you for your attention to this important matter. I look forward to your leadership in making [State] a model for privacy and digital protection.

Sincerely,

[Your Name]

FEDERAL LEVEL

While we've already gone over the lack of urgency to pass regulations at the federal level, I think it would be helpful to hold our legislators to the fire and ask them to stop bending to lobbyists.

Let's ask them to put the needs of their constituents ahead of lobbying dollars.

Letter to Your Federal Senator or Congressperson

Here's a start. Make it your own by adding the issue you think is most important.

[Your Name]
[Your Address]
[City, State, ZIP]
[Email Address]
[Phone Number]
[Date]

The Honorable [Full Name]
[Title, e.g., Senator/Representative]
[Office Address]
[City, State, ZIP]

Dear Senator/Representative [Last Name],

I am writing to express my deep concern about the influence of Big Tech lobbying on policy decisions that affect the privacy, safety, and well-being of American citizens, especially our children. As a constituent, I urge you to take decisive action to protect the American people from harmful data practices, invasive surveillance, and addictive algorithms that prioritize profit over people.

Tech companies are gathering vast amounts of personal data on your constituents and providing no transparency about its usage and purposes, manipulating user behavior with addictive algorithms, and creating systems that exploit children's vulnerability. Unfortunately, lobbying by these companies has played a role in preventing meaningful reforms that could protect citizens' privacy and digital rights. As our elected representative, I ask that you work to sever the financial influence of Big Tech and support strong regulatory measures in the following areas:

PRIVACY PROTECTIONS: Americans deserve control over their personal data. Laws like the **American Data Privacy Protection Act (ADPPA)** (H.R. 8152) and the

Consumer Data Protection Act (CDPA) (S. 2737) should be prioritized to enforce comprehensive privacy standards for tech companies.

DATA COLLECTION TRANSPARENCY: We need transparency in how companies collect, store, and sell our data. Proposals such as the **Data Accountability and Transparency Act** (S. 447) would help provide consumers with clear, accessible information about data practices.

ADDICTIVE ALGORITHMS: Companies often design algorithms that prioritize user engagement at the cost of mental health and well-being. I encourage you to support bills like **The Children and Media Research Advancement Act** (S. 1163), which calls for research into the effects of digital platforms on children, and to explore legislation that limits these manipulative techniques.

AI AND AUTOMATED DECISION-MAKING: With the rapid rise of AI, it's critical that we establish regulations to ensure AI systems are transparent, explainable, and accountable. Bills like the **Algorithmic Accountability Act** (S. 2292) should be revisited to ensure fairness and transparency in automated decision-making systems.

AI LABELING: Clear labeling of AI-generated content is essential to combat misinformation and prevent manipulation. I urge you to support initiatives that require AI labeling, such as those outlined in the **Artificial Intelligence Accountability Act** (H.R. 4855).

> While bills like these have been introduced, they have not progressed due to heavy opposition from powerful tech lobbying groups. By supporting these measures and cutting ties with tech industry lobbying, you can help restore trust in our government and ensure that the technology we use serves the people, not corporate interests.
>
> Thank you for your time and consideration. I hope to see you champion privacy protections, transparency, and ethical AI practices that safeguard our rights and the future of our children.
>
> Sincerely,
>
> [Your Name]

THE GLOBAL LEVEL

While America's love for capitalism has turned data collection into a billion-dollar industry and supports the use of AI without regulation, the sentiment is not shared globally. The United Nations released some principles and guidelines that reflect global consensus on issues such as data privacy, artificial intelligence, and human rights. These are intended to serve as frameworks for countries to follow when creating their own laws.

They've released several guidelines through the UNESCO Recommendation on the Ethics of Artificial Intelligence and the UN Secretary-General's Roadmap for Digital Cooperation. The ones that I find most important are these:

- **HUMAN RIGHTS AND DIGNITY:** AI should be developed and used in ways that protect and promote human rights and human dignity. This includes ensuring that AI systems do not violate individuals' right to privacy, their right to freedom of expression, or their right to non-discrimination.

- **PRIVACY AND DATA PROTECTION:** The recommendation calls for AI systems to respect privacy rights by ensuring that data collection and processing are done transparently, with proper informed consent from individuals. Data protection is emphasized, and it is stressed that AI should not be used to violate privacy through excessive surveillance or the unwarranted collection of personal data.

- **TRANSPARENCY:** AI systems should be transparent in their operation, meaning that individuals should have a clear understanding of how their data is being used and how AI decisions are made. This transparency helps protect privacy and enables people to make informed choices.

- **FREEDOM FROM ARBITRARY SURVEILLANCE:** There should be strict limits on surveillance technologies, particularly AI-driven surveillance systems like facial recognition. Governments are urged to ensure that these technologies are not used arbitrarily or excessively, as they can easily infringe on privacy and civil liberties.

- **PROTECTION OF PERSONAL DATA:** The Human Rights Committee encourages governments to adopt legal frameworks that protect personal data from unauthorized use and ensure that individuals have control over their data. This includes access to their data, the ability to correct it, and the right to have it erased.

- **DATA RETENTION:** The Human Rights Committee advocates for strict rules on how long personal data can be stored and for what purposes, ensuring that data retention policies are clear and that data is not held longer than necessary.

While these guidelines are voluntary, they do provide an international standard that governments are encouraged to adhere to. Europe has done so, adopting the General Data Protection Regulation (GDPR). The GDPR fines companies 4 percent of their global revenue if they don't comply with the various protective requirements, including allowing people to see what data is collected, providing an understanding of what's happening with it, and letting them erase it. Plus, the regulation will not allow collection of data without a consumer's consent.

In early 2025, US Vice President JD Vance told heads of state and CEOs gathered in Paris for the Artificial Intelligence Action Summit, "We believe that excessive regulation of the AI sector could kill a transformative industry just as it's taking off."

European legislators and those at the UN do not agree. In 2024, the European Parliament approved the world's first comprehensive framework for constraining the risks of artificial intelligence (AI). According to the BBC, "The main idea of the law is to regulate AI based on its capacity to cause harm to society. The higher the risk, the stricter the rules. AI applications that pose a 'clear risk to fundamental rights' will be banned, for example some of those that involve the processing of biometric data. AI systems considered 'high-risk,' such as those used in critical infrastructure, education, healthcare, law enforcement, border management, or elections, will have to comply with strict requirements."[2]

It's helpful to know that while the United States doesn't agree to such regulations, other countries are working hard to protect humanity. While the UN doesn't have the capacity to "punish" a country that doesn't follow its global guidelines, the US might face increasing international pressure to comply with global standards. This could affect trade relations, diplomatic ties, or access to global markets, especially if other countries feel that the US is not following best practices for AI or data privacy.

If you wanted to file a complaint that a certain country is violating human rights by collecting personal data without obvious consent, you could raise a flag on their twenty-four-hour hotline (+1 (212) 963-1111) or report wrongdoing to their office of oversight at www.oios.un.org.

TRANSPARENCY

If there is one issue to really fight for, I would say it's transparency. Sure, we can't regulate the actions of the Big Tech companies, but we can at least get them to tell us what they're doing. I want transparency. I say we make this our big push.

The lack of transparency makes me feel completely out of control. They know so much about me, but I don't know what that means. This could potentially be a violation of a human right.

We want to know what information about us they're collecting. We want to know what kinds of algorithms they're enacting, and we want the option to be able to make the decision to use or not use their product if we don't agree with what they reveal.

Requiring transparency would allow for better decision-making. For example, you can choose between the way Garmin and Apple watches use your biometric data—and enter into the agreement that feels safer to you.

You might conduct your searches with Bing instead of Google once you learn all the ways Google is collecting your search history data. This knowledge would allow us to compare the personal benefits and the data cost of that product or service. If you click on an Instagram ad and plug in your email, does that mean your email will be used on a newsletter list *and* sold to other, similar vendors *and then* sold to data banks to be used in perpetuity?

This clarity would help us choose companies based on data usage practices that are more closely aligned to our values. Withholding the way they use our information deprives us of the right to make choices reflecting our personal values, beliefs, and priorities. Big Tech is infringing on an inherent human right to control the information that defines our lives, interests, and identities.

Have you ever been to a parking garage in California? Or tried to buy a piece of furniture from Amazon and have it sent to your summer home in California? If so, you might have noticed a Prop 65 warning. Prop 65 requires businesses to provide warnings about chemicals in their products that cause cancer, birth defects, or other reproductive harm, allowing California citizens to make more informed decisions. The list of chemicals that must be noted on labels contains a wide range of naturally occurring and synthetic chemicals that include additives, pesticides, food, drugs, dyes, or solvents that are commonly used in household and industrial products. Do I want to use a table that has formaldehyde in it as my primary dining table every night when it might cause me reproductive harm? Probably not. Do I want to park daily in a garage whose lack of ventilation allows me to breathe in too much carbon monoxide? Nope.

The proposition doesn't prohibit the use of these chemicals. It simply allows us all to make smarter choices. I would like to see a similar risk factor label on every AI-generated product or AI-generated social media content. Sometimes you're at risk of learning misinformation. Sometimes you're at risk of falling for a scam. Sometimes you're at risk of being led toward eating disorders or suicidal thoughts. This needs to be explicit when you log on to Instagram, Facebook, or clickbait.

Transparency in AI is just as imperative. We might read or watch AI-generated content and use the information it provides us to make important decisions, make purchases, learn new information, or vote! That information may be incorrect. This could affect our lives and our mental health more than, say, eating a protein bar. Yet the bar manufacturer must list its ingredients, its calorie, sugar, fiber, protein, sodium, and fat content. Every item that employs AI should list data use, data storage, security and privacy practices, and algorithms.

If they won't label the risk factors, what if they were simply required to disclose the algorithms they use to prioritize content and personalize experiences, especially when these algorithms impact vulnerable users like children. This would include showing how data is used to manipulate user behavior or create addiction. An independent third-party audit of algorithms and data practices should be mandated to ensure they don't disproportionately harm certain groups (e.g., children, vulnerable individuals, particular ethnic groups) or promote harmful behavior (e.g., cyberbullying, hate speech).

If you pick one issue, I think transparency is it. So take that to the state level, urge companies in your area to label, and make a lot of noise until some kinds of transparency laws are enacted.

WHAT ELSE

Okay, so you've talked to your state representative and the school board, and you've gotten phones banned from your kids' classrooms. Great! What else can you do?

There are ways you can help yourself be less lonely, broke, and powerless. While I have described the actions of Big Tech and data brokers as predatory, it doesn't mean we have to be prey. We can still take action while we wait on changes and regulations.

Remember: We are Big Tech's biggest asset. They would never make a dime without our attention, our content, and our dollars. The best thing we can do for ourselves and for the good of everyone is to log off.

In the beginning of 2025, shortly after President Trump took office for the second time, Target rolled back its Diversity, Equity, and Inclusion programs. This was a slap in the face for Black shoppers who spend $12 million at the retail giant daily. Immediately angry shoppers of all races began a boycott, and it worked! Target's share price dropped 18 percent, and their first quarterly report was down 3 percent.[3]

Boycotts can work and effect change. The first step is to unsubscribe and unplug. This might seem obvious, but it's painfully hard for most of us. With all the intentional design elements, dings, and algorithms inviting you back, it takes a lot of work to put it all to the side for even a few hours a day.

It helps me to remember that the big billionaires want us to be hooked. We are millions of people. Here's a novel idea. Let's just quit. Let's stage a walkout, a shutdown, a social media cleanse.

If we all shut down our computers for the day, if we didn't engage on social media, click the bait, let ourselves be led from ad to website like Pavlov's dogs by the bell—and perhaps more important if we didn't buy anything online—what would they do? What if we went on a social strike until Big Tech changed some of its policies? We could give up social media for a day. A week. For as long as necessary.

The lifeblood of many social media platforms are the advertisements they can place based on your data. If you're not there, they can't earn money from advertisers who want your eyeballs to see their information. And if you stop scrolling, they can't collect information on you to sell or use for more ads. Your lack of participation takes away two streams of income for them.

What would happen if brand strategists could not gauge how many people clicked on their products? Data harvesting would cease, the data brokers would halt, and the transactional chain of events that drive people's data toward advertisements designed to get them to click "buy now" in their shopping carts would be disrupted.

If we all did it, Big Tech might have to say: Okay! We will enact safer regulations. We will care about your experience. We will stop marketing to kids. We will label when we use AI. We will tell you what we're doing with your data.

At least that's my hope.

Another pleasant side effect of this strike might be the healing of humans. People will turn to each other, as they do during power outages or in the aftermath of hurricanes, and wait together and talk. Unexpected connections may occur. Teens may stop looking toward Insta for validation for one day and go visit their friends. I might look up from my emails and accept a dinner invitation.

The thing is, I don't have a large enough following to be the Greta Thunberg of a social media strike. I did pay a PR company at one point to run an Instagram account for me so that I could start telling people about Lotic. It didn't last long. I'm not meant for Insta. So, it's up to you. Please, can you be the Greta Thunberg of a social media strike?

If you aren't ready to organize a strike, go on a little social media cleanse. Cleanse with your friends. Convince your friends to unplug and go to lunch every Friday. Every little bit helps. Just twenty people can create a lot of data for Big Tech, and losing small groups of friends here and there means a lot of information lost.

BUILD YOUR MINDFUL SPACE

How else can you gain back a bit of the agency you've lost? Get rid of as many apps as possible on your phone. Who knows how many are tracking your activity, location, and other vital statistics? Get them off! You barely use them anyway.

Get mindful about your tech usage. Before you pick up your phone, ask yourself why you're doing it. Are you really looking for something, or just mindlessly scrolling? Acknowledge whether or not mindlessly scrolling helps. Many of us pick up our phones to take a break, to let our minds rest. Does your mind really rest while scrolling? My mind doesn't rest when I am mindlessly scrolling. It makes me feel anxious and restless! Asking yourself how you feel about that scrolling session can develop an awareness to help you curb unnecessary phone use.

Change your home screen. One of the best tools I have used to help me is changing my phone's wallpaper to just a plain color and moving all the apps to the second page. When I look at my phone, I see only a black square. There is no longer any danger of getting pulled into another app and losing my concentration for an accidental hour. If I look at my phone, I must type in the app I want to use, and it requires intention. This trick has truly changed my entire experience.

Alternatively, you can switch to a simpler phone. Some people find that moving to a basic phone (often called a "dumb phone") or a phone with limited functionality helps them break free from constant phone use. These phones are centered on calling and texting, minimizing distractions.

You can also set some rules for yourself or your family at home:

- **NO PHONES IN THE BEDROOM:** Try charging your phone in a different room to avoid the temptation of checking it first thing in the morning or late at night. This also helps improve your sleep quality.
- **NO PHONES DURING MEALS:** Keep your phone away from the dining table to focus on the people around you and enjoy your meals without distractions.
- **DESIGNATE ZONES FOR FOCUS:** Create areas in your home or office that are "phone-free zones" to help you focus better on the task at hand, whether it's working or relaxing.

Make it a part of your self-care to unsubscribe. Every day, find a brand that sends you emails and makes you want things you don't need, and tell them to stop sending you information. You don't need it. When you want a pair of shoes, you can look into shoes. You don't need shoes in your inbox.

If you're still on social media, follow the people talking about decluttering and sustainability and minimalism and underconsumption. Learn how good it feels to have less.

Stop doing the least. Curb some of that technology-induced loneliness with a small bit of extra effort. This was hard for me because it sure is easier and more comfortable to stay in your PJs and scroll. Remember that each time you scroll for that lost hour in your day, you're doing just what Big Tech wants. Take an in-person class and learn something you've always wanted to learn! Ceramics, painting, motocross. Whatever. When you do something that fulfills your passion, you'll meet people with similar passions.

Record your stories. Whether you use Lotic or your voice notes, tell your story so you can change the wiring of your brain and get clear on what kinds of thoughts you might be unknowingly holding inside. You might just learn something or see some biases that have been controlling your life up until now. They may lose some of their power. I wish you many realizations.

Build community. This might be the most important piece. Start a book club in the park. Start a repair club on the weekends where you fix people's old things so they don't have to buy more. (One of my favorite parts about being a father is having to devise material solutions to fix broken things, and, around kids, something is always needing repair.) Set a goal to hang out with a different friend every Friday. Talk to your local coffee shop about starting a no-technology day.

Go out into nature. Tell a joke. Volunteer. Send someone a text that says you love them. Try to do some good in the world. The more you do in real life, the less you will need to think about your phone.

If you have the energy, run for office. We need candidates that will not bow down to the billionaires. We need people in office who prioritize tech regulation and care about this stuff.

If you're a coder, an inventor, a tech enthusiast, start trying to beat me. Make your own version of my idea. Create a marketplace that uses stories as currency. I beg you! I want to see it implemented in my lifetime, whether I do it or not. My email is open: ww4@lotic.ai.

Whatever you do, never forget the power of your unique story, whether it's used as currency or not. What you have been through is important. It holds so much information about who you really are, and it deserves your attention. Value your own story and the stories of others. Learn from them. Cherish them. Realize they always lead to connection. Don't ever stop telling them.

We can't fix everything. No society will be perfect. Yet, we have options. We don't have to be stuck with what we've been given. Ever. We don't have to accept a life that isn't working for us anymore. We can actually do something to help ourselves and help the world. And if we can't do that, at least we can try.

That's what I want to tell my kids I did. I hope you'll do the same.

APPENDIX

Pick Your Misperception

Fifty common biases that may (unknowingly) drive our stories, identified by Amos Tversky and Daniel Kahneman.

1. Ostrich Effect
DEFINITION: The tendency to avoid information or situations that are perceived as negative or uncomfortable, like an ostrich burying its head in the sand.
EXAMPLE: Imagine a person who knows their financial situation is precarious but chooses not to look at their bank account or credit card statements. They avoid checking because they fear seeing the negative balance or realizing how much they owe, opting to stay in the dark instead.

2. Bandwagon Effect
DEFINITION: The tendency to adopt certain behaviors, beliefs, or opinions because others are engaging in these behaviors, beliefs, or opinions.
EXAMPLE: You start watching a new TV show just because all your friends are talking about it, even if you're not initially interested.

3. The Peak-End Rule
DEFINITION: The tendency for people to judge an experience based largely on how they felt at its peak (the most intense point) and its end, rather than the overall experience.

EXAMPLE: After going on a vacation, someone might rate the entire trip as excellent or terrible based only on one incredible moment (e.g., watching a beautiful sunset) and how it ended (e.g., missing a flight or a last-minute mishap).

4. The Forer Effect (aka Barnum Effect)
DEFINITION: The tendency for people to accept vague or general statements as personally meaningful or accurate, even when the statements apply to a wide range of people. This bias is often used in astrology, personality tests, and horoscopes.
EXAMPLE: After reading a horoscope that says, "You may face challenges today, but you will overcome them," someone might feel that the prediction speaks directly to their life.

5. Hindsight Bias
DEFINITION: The inclination to see events as having been predictable after they have already occurred.
EXAMPLE: After a sports team loses, you might think, "I knew they were going to lose," even though the outcome was uncertain beforehand.

6. Self-Serving Bias
DEFINITION: The tendency to attribute positive outcomes to one's own abilities and negative outcomes to external factors.
EXAMPLE: If you ace a test, you attribute it to your intelligence, but if you fail, you blame it on the difficulty of the test.

7. Status Quo Bias
DEFINITION: The preference for things to remain the same rather than change.
EXAMPLE: Sticking with an old phone because it's familiar, even though newer models have better features.

8. Overconfidence Bias
DEFINITION: The tendency to overestimate one's abilities, knowledge, or predictions.

EXAMPLE: A person might think they are the best driver on the road, even though they have a history of small accidents.

9. Framing Effect
DEFINITION: The way information is presented can affect decision-making and judgment.
EXAMPLE: People are more likely to choose a surgery if it has a "90 percent success rate" versus a "10 percent failure rate," even though both statements are identical.

10. Attribution Bias
DEFINITION: The tendency to attribute one's own actions to external factors, while attributing others' actions to their character or personality.
EXAMPLE: You fail a test and blame it on the teacher being unfair, but when someone else fails, you think it's because they didn't study enough.

11. Planning Fallacy
DEFINITION: The tendency to underestimate the time, costs, and risks of future actions, while overestimating the benefits.
EXAMPLE: Planning a home renovation and underestimating how much time and money it will actually take.

12. Sunk Cost Fallacy
DEFINITION: The tendency to continue an endeavor once an investment of time, money, or effort has been made, even if it no longer makes sense.
EXAMPLE: Continuing to watch a movie you're not enjoying because you've already watched an hour of it.

13. Negativity Bias
DEFINITION: The tendency to give more weight to negative experiences or information than positive ones.
EXAMPLE: After receiving five positive reviews and one negative review, you focus more on the negative feedback.

14. Optimism Bias
DEFINITION: The belief that things are more likely to go well than they actually are.
EXAMPLE: A person may think they're unlikely to get into a car accident because they believe it won't happen to them.

15. In-Group Bias
DEFINITION: The tendency to favor people who belong to your own group over those in other groups.
EXAMPLE: Supporting a sports team simply because they're from your city, even if they're not performing well.

16. Out-Group Homogeneity Bias
DEFINITION: The tendency to view members of an outcast group as being more similar to each other than they actually are.
EXAMPLE: Thinking all people from a different country have the same culture or worldview.

17. Halo Effect
DEFINITION: The tendency to let an overall impression of a person influence specific judgments about them.
EXAMPLE: If someone is physically attractive, you might assume they are also intelligent or kind.

18. Horn Effect
DEFINITION: The opposite of the Halo Effect, where a negative impression of someone influences other perceptions of them.
EXAMPLE: If someone is rude to you once, you might assume they are generally unpleasant in all situations.

19. False Consensus Effect
DEFINITION: The tendency to overestimate the extent to which others share your beliefs and behaviors.
EXAMPLE: You might assume everyone agrees with your political views simply because they live in your neighborhood or have similar friends.

20. Illusion of Control
DEFINITION: The tendency to believe that one has control over situations that are actually beyond one's control.
EXAMPLE: Thinking you can influence the outcome of a dice roll because you're "lucky."

21. Dunning-Kruger Effect
DEFINITION: The phenomenon where people with low ability to perform a task overestimate their ability to perform that task.
EXAMPLE: A beginner chess player who won a few games online thinks they can beat a grandmaster.

22. Endowment Effect
DEFINITION: The tendency to assign more value to things merely because you own them.
EXAMPLE: You're reluctant to sell a secondhand item for less than its original price because you feel it's worth more now that it's yours.

23. Choice-Supportive Bias
DEFINITION: The tendency to retroactively ascribe positive attributes to choices one has made, even if the decision wasn't the best.
EXAMPLE: After buying an expensive car, you convince yourself it was a great decision, despite the car's plethora of mechanical issues.

24. Recency Effect
DEFINITION: The tendency to give undue weight to the most recent information you've heard or seen.
EXAMPLE: Remembering the last point made during a presentation more vividly than the earlier points.

25. Primacy Effect
DEFINITION: The tendency to better remember information that was presented first.
EXAMPLE: In a list of words, you're more likely to remember the first few words than the middle ones.

26. Cognitive Dissonance

DEFINITION: The mental discomfort experienced when holding two or more conflicting beliefs, values, or attitudes.

EXAMPLE: A smoker justifies their habit by claiming, "I'm not addicted; I just enjoy it," even though they know it's harmful.

27. Mere Exposure Effect

DEFINITION: The tendency to develop a preference for things merely because they are familiar.

EXAMPLE: You start liking a song after hearing it repeatedly, even if you didn't like it at first.

28. Clustering Illusion

DEFINITION: The tendency to see patterns in random data.

EXAMPLE: Believing a series of coin flips results in a pattern like "heads-tails-heads," even though it's purely random.

29. Just-World Hypothesis

DEFINITION: The belief that people get what they deserve, leading to the tendency to blame victims for their misfortunes.

EXAMPLE: Thinking a person who loses their job must have done something wrong, rather than considering external factors that may have contributed to that loss.

30. Misinformation Effect

DEFINITION: The distortion of memory caused by exposure to incorrect or misleading information.

EXAMPLE: After watching a news story about a politician, you might decide to vote for them when you're not sure if that news story was even credible.

31. False Memory

DEFINITION: The tendency to recall something inaccurately or as a memory of an event that didn't occur.

EXAMPLE: Believing you met a celebrity when you didn't, because you'd remembered a friend talking about her experience meeting a celebrity.

32. Illusory Superiority
DEFINITION: The tendency to view oneself as better than others, especially in areas of ability or achievement.
EXAMPLE: Believing you're a better driver than the average person, even if you're not.

33. Bystander Effect
DEFINITION: The phenomenon where people are less likely to help in an emergency when others are present.
EXAMPLE: Witnessing an accident and assuming someone else will call for help, so you don't take action.

34. Actor-Observer Bias
DEFINITION: The tendency to attribute your own actions to external factors, while attributing others' actions to their personality or character.
EXAMPLE: If you're late, you blame traffic; if someone else is late, you think they're irresponsible.

35. Belief Bias
DEFINITION: The tendency to judge arguments based on the believability of their conclusion, rather than on the strength of the arguments.
EXAMPLE: If you already believe that climate change is a hoax, you're more likely to accept weak arguments that support this belief.

36. The False Uniqueness Effect
DEFINITION: The tendency to underestimate how common our own positive behaviors or traits are, leading us to believe we are more unique than we actually are.
EXAMPLE: A person who volunteers regularly might assume they are exceptional for doing so, even though many others also contribute to charitable causes.

37. Empathy Gap
DEFINITION: The inability to properly understand or share the feelings of others, often due to a lack of personal experience.
EXAMPLE: A person with no experience of chronic illness may have difficulty understanding the struggles of someone who is ill.

38. Affect Heuristic
DEFINITION: Making decisions based on emotions or feelings, rather than logical reasoning.
EXAMPLE: You choose a product because you have a positive emotional association with its brand, even if it's not the best option.

39. Regret Aversion
DEFINITION: The tendency to avoid decisions that could lead to regret, even if those opportunities could lead to great benefits.
EXAMPLE: You avoid making an investment because you fear it might go wrong, and you will regret it later.

40. Social Proof
DEFINITION: The tendency to assume that the actions of others reflect correct behavior, especially in uncertain situations.
EXAMPLE: You're more likely to buy a product if you see that it has many positive reviews or it is popular.

41. Zero-Risk Bias
DEFINITION: The preference for reducing a small risk to zero, even if larger risks remain.
EXAMPLE: Implementing a costly policy to eliminate one small risk, even if it doesn't address much larger, more important risks.

42. Conservatism Bias
DEFINITION: The tendency to favor prior evidence over new evidence or new information.
EXAMPLE: A person dismisses recent research that contradicts their long-held beliefs because they prefer the old information.

43. Illusion of Transparency
DEFINITION: The tendency to overestimate how well others can understand our thoughts, feelings, or emotions.
EXAMPLE: You think everyone knows you're nervous during a presentation because you feel nervous, but others may not notice.

44. Paradox of Choice
DEFINITION: The phenomenon where having too many options leads to less satisfaction or more difficulty making a decision.
EXAMPLE: Feeling overwhelmed by a large multi-page menu at a restaurant and later leaving the restaurant unsure about whether you made the best choice.

45. Temporal Discounting
DEFINITION: The tendency to value immediate rewards more highly than future ones.
EXAMPLE: Choosing to spend money on something fun now rather than saving for a more important goal later.

46. Gambler's Fallacy
DEFINITION: The belief that future probabilities are influenced by past events in a random sequence.
EXAMPLE: Believing a roulette wheel is "due" to land on red after several blacks, even though each spin is independent.

47. Pari Passu Bias
DEFINITION: The tendency to assume that the experience of one group in a situation applies equally to other groups.
EXAMPLE: Assuming that all immigrants face the same challenges in adapting to a new country, even though each group's experience can vary.

48. Hot-Hand Fallacy
DEFINITION: The belief that a person who has experienced success has a higher chance of success in subsequent attempts.

EXAMPLE: Believing a basketball player is "on fire" after a few successful shots and betting they'll continue to make shots.

49. Curse of Knowledge
DEFINITION: The difficulty that experts have in imagining what it's like for someone who is less knowledgeable about a subject.
EXAMPLE: A doctor explaining a complex medical procedure to a patient might use jargon that the patient doesn't understand, assuming the patient knows what they mean.

50. The Spotlight Effect
DEFINITION: The tendency to overestimate how much other people notice your appearance, behavior, or mistakes.
EXAMPLE: You trip in public and feel like everyone is staring at you, even though most people didn't notice.

NOTES

Introduction
1. Josh Howarth, "7 Top Gen Z Trends for 2024," Exploding Topics, Updated April 23, 2025, https://explodingtopics.com/blog/gen-z-trends.

Chapter 2
1. Samantha Murphy Kelly, "Elon Musk Says AI Will Take All Our Jobs," CNN Business, May 23, 2024, https://edition.cnn.com/2024/05/23/tech/elon-musk-ai-your-job/index.html.
2. Julia Corcoran, Peter O'Dowd, and Jill Ryan, "Is the American Way of Socializing Going Down?" *Here & Now*, WBUR, February 19, 2024, https://www.wbur.org/hereandnow/2024/02/19/american-socialization-down.
3. Rhitu Chatterjee, "Teen Girls and LGBTQ+ Youth 'Plagued by Violence and Trauma,' Survey Says," NPR, February 13, 2023, https://www.npr.org/sections/health-shots/2023/02/13/1156663966/teen-girls-and-lgbtq-youth-plagued-by-violence-and-trauma-survey-says.
4. Derek Thompson, "Why American Teens Are So Sad," *The Atlantic*, April 11, 2022, https://www.theatlantic.com/newsletters/archive/2022/04/american-teens-sadness-depression-anxiety/629524/.

Chapter 3
1. Office of the Surgeon General, "Our Epidemic of Loneliness and Isolation: The U.S. Surgeon General's Advisory on the Healing Effects of Social Connection and Community," US Department of Health and Human Services, 2023, https://www.hhs.gov/sites/default/files/surgeon-general-social-connection-advisory.pdf.

2. Hamish M. E. Foster et al., "Social Connection and Mortality in UK Biobank: A Prospective Cohort Analysis," *BMC Medicine* 21 (November 10, 2023): 384, https://doi.org/10.1186/s12916-023-03055-7.

3. Milena Batanova, Richard Weissbourd, and Joseph McIntyre, *Loneliness in America: Just the Tip of the Iceberg?* (Cambridge, MA: Making Caring Common, Harvard Graduate School of Education, October 2024), https://mcc.gse.harvard.edu/reports/loneliness-in-america-2024.

4. Julia Corcoran, Peter O'Dowd, and Jill Ryan, "Is the American Way of Socializing Going Down?" *Here & Now*, WBUR, February 19, 2024, https://www.wbur.org/hereandnow/2024/02/19/american-socialization-down.

5. Cassandra Stone, "TikTok Nails the Absence of Boomer-Age Grandparents," Motherly, September 1, 2023, https://www.mother.ly/news/viral-trending/boomer-grandparents-tiktok/.

6. Marc Dunkelman, "The Transformation of American Community," *National Affairs*, no. 63 (Spring 2025), https://www.nationalaffairs.com/publications/detail/the-transformation-of-american-community.

7. Decca Aitkenhead, "What Happened When I Made My Sons and Their Friends Go Without Smartphones," *The Times*, February 20, 2025, https://www.thetimes.com/life-style/parenting/article/teenage-sons-friends-without-smartphones-experiment-vpcnbj58d.

Chapter 4

1. "The State of U.S. Household Wealth," Federal Reserve Bank of St. Louis, accessed April 30, 2025, https://www.stlouisfed.org/community-development/publications/the-state-of-us-household-wealth.

2. Katie Reilly, "Exactly How Teachers Came to Be So Underpaid in America," *TIME*, September 13, 2018, https://time.com/longform/teaching-in-america/.

3. National Center for Education Statistics, Digest of Education Statistics 2021, Table 211.50, "Estimated Average Annual Salary of Teachers in Public Elementary and Secondary Schools; Selected School Years, 1959–60 Through 2021–22," accessed April 30, 2025, https://nces.ed.gov/programs/digest/d21/tables/dt21_211.50.asp.

4. Bernie Sanders, "The Rich-Poor Gap in America Is Obscene. So Let's Fix It—Here's How," Senator Bernie Sanders, March 29, 2021, https://www.sanders.senate.gov/op-eds/the-rich-poor-gap-in-america-is-obscene-so-lets-fix-it-heres-how/.

5. Aliya Uteuova, "'Don't Be Scared of Beans': How Readers Are Handling US Grocery Inflation," *The Guardian*, August 28, 2024, https://www.theguardian.com/environment/article/2024/aug/28/inflation-groceries-tips.

6. Zach Hrynowski and Stephanie Marken, "Gen Z Voices Lackluster Trust in Major U.S. Institutions," Gallup News, September 14, 2023, https://news.gallup.com/opinion/gallup/510395/gen-voices-lackluster-trust-major-institutions.aspx.

7. Katy Marquardt, "What's the Average Amount of Credit Card Debt in the U.S." *U.S. News & World Report*, October 2024, https://money.usnews.com/credit-cards/articles/whats-the-average-amount-of-credit-card-debt-in-the-u-s#.

8. Senate Hearing on Competition in the Credit Card Market, Senate Judiciary Committee, November 19, 2024, https://www.c-span.org/program/senate-committee/senate-hearing-on-competition-in-the-credit-card-market/652315.

9. Scott I. Rick, Beatriz Pereira, and Katherine A. Burson, "The Benefits of Retail Therapy: Making Purchase Decisions Reduces Residual Sadness," Ross School of Business Working Paper No. 1208, University of Michigan, January 2014, https://deepblue.lib.umich.edu/bitstream/handle/2027.42/100258/1208_Rick_Jan14.pdf.

10. Jennifer Flanagan, "Social Media Solidifies Its Role as a Leading Marketplace and Consumer Influencer," Adtaxi, October 9, 2024, https://www.adtaxi.com/blog/social-media-solidifies-its-role/.

11. "How Many Orders Does Amazon Get & Deliver per Day?" Capital One Shopping, last modified April 15, 2025, https://capitaloneshopping.com/research/amazon-orders-per-day/#:~:text=Highlights.,of%25200ver%252520600%252520million%252520products.

12. Chavie Lieber, "Tech Companies Use 'Persuasive Design' to Get Us Hooked. Psychologists Say It's Unethical," *Vox*, August 8, 2018, https://www.vox.com/2018/8/8/17664580/persuasive-technology-psychology.

Chapter 5

1. Kelsey Weekman, "How 'Get Ready with Me' Videos Became a Social Media Trend That Won't Go Away," Yahoo News, December 20, 2023, https://www.yahoo.com/news/how-get-ready-with-me-videos-became-a-social-media-trend-that-wont-go-away-195309611.html.

2. Matthew Bremner, "How to Be Human: The Man Who Was Raised by Wolves," *The Guardian*, August 28, 2018, https://www.theguardian.com/news/2018/aug/28/how-to-be-human-the-man-who-was-raised-by-wolves.

3. GO-Globe, "Every 60 Seconds What Happens on the Internet in 2025?" GO-Globe, April 2025, https://www.go-globe.com/things-that-happen-on-internet-every-60-seconds; Susie Marino, "What Happens in an Internet Minute: 90+ Fascinating Online Stats," LocaliQ, January 17, 2025, https://localiq.com/blog/what-happens-in-an-internet-minute/.

4. "Social Media, Suicidal Thoughts and an Identity Crisis Among Young Adults," *U.S. News & World Report*, September 29, 2023, https://www.usnews.com/news/health-news/articles/2023-09-29/social-media-suicidal-thoughts-and-an-identity-crisis-among-young-adults.

5. Phil Borges (Crazywise), "Gabor Maté – Authenticity vs. Attachment," YouTube video, 4:18. May 14, 2019, https://www.youtube.com/watch?v=l3bynimi8HQ&t=138s.

Chapter 6

1. Trevor Haynes, "Dopamine, Smartphones & You: A Battle for Your Time," Science in the News, Harvard University, May 1, 2018, https://unplugged.sunygeneseoenglish.org/wp-content/uploads/sites/31/2019/11/Domamine-PDF.pdf.

2. Martin Korte, "The Impact of the Digital Revolution on Human Brain and Behavior: Where Do We Stand?" *Dialogues in Clinical Neuroscience* 22, no. 2 (June 2020): 101–111, https://pmc.ncbi.nlm.nih.gov/articles/PMC7366944/.

3. Qinghua He, Ofir Turel, and Antoine Bechara, "Brain Anatomy Alterations Associated with Social Networking Site (SNS) Addiction," *Scientific Reports* 7 (March 23, 2017): Article 45064, https://pmc.ncbi.nlm.nih.gov/articles/PMC5362930/.

4. Julio Vincent Gambuto, *Please Unsubscribe, Thanks: How to Take Back Our Time, Attention, and Purpose in a World Designed to Bury Us in Bullsh*t* (New York: Atria Books, 2023).

5. Martin Korte, "The Impact of the Digital Revolution on Human Brain and Behavior: Where Do We Stand?" *Dialogues in Clinical Neuroscience* 22, no. 2 (June 2020): 101–111, https://pmc.ncbi.nlm.nih.gov/articles/PMC7366944/.

6. Arkady Yerukhimovich et al., "Can Smartphones and Privacy Coexist? Assessing Technologies and Regulations Protecting Personal Data on Android and iOS Devices," RR-1393-DARPA (Santa Monica, CA: RAND Corporation, October 27, 2016), https://www.rand.org/pubs/research_reports/RR1393.html.

7. "Data Broker Market Demand, Research Insights by 2031," Transparency Market Research, last modified April 2025, https://www.transparencymarketresearch.com/data-brokers-market.html.

8. Kashmir Hill, "How Target Figured Out a Teen Girl Was Pregnant Before Her Father Did," *Forbes*, February 16, 2012.

9. "Federal Agencies Use Cellphone Location Data for Immigration Enforcement," *Wall Street Journal*, May 10, 2019, https://www.wsj.com/articles/federal-agencies-use-cellphone-location-data-for-immigration-enforcement-11581078600.

10. Caroline Haskins, "A Tech Company Spied on Police Brutality Protesters," BuzzFeed News, June 25, 2020, https://www.buzzfeednews.com/article/carolinehaskins1/protests-tech-company-spying.

11. Donie O'Sullivan, "How the Cell Phones of Spring Breakers Who Flouted Coronavirus Warnings Were Tracked," CNN, April 4, 2020, https://edition.cnn.com/2020/04/04/tech/location-tracking-florida-coronavirus/index.html.

12. "Apple vs Meta Threads: The Illusion of Privacy," Growth.Design, https://growth.design/case-studies/apple-privacy-policy.

13. Kari Santos, "Giving Google a Piece of Your Mind," *Daily Journal*, February 2, 2014, https://www.dailyjournal.com/articles/291469-giving-google-a-piece-of-your-mind.

14. Nick Statt, "Google Will Stop Scanning Your Gmail Messages to Sell Targeted Ads," The Verge, June 23, 2017, https://www.theverge.com/2017/6/23/15862492/google-gmail-advertising-targeting-privacy-cloud-business.

15. Lily Hay Newman, "Think Twice Before Using Facebook, Google, or Apple to Sign In Everywhere," *Wired*, September 21, 2020, https://www.wired.com/story/single-sign-on-facebook-google-apple/.

Chapter 7

1. Mustafa Suleyman and Michael Bhaskar, *The Coming Wave: Technology, Power, and the Twenty-First Century's Greatest Dilemma* (New York: Crown, 2023).
2. Rebecca Voelker and Yulin Hswen. "Can Predictive AI Improve Early Detection of Sepsis and Other Conditions?" *JAMA* 330, no. 20 (November 28, 2023): 1939–42, https://doi.org/10.1001/jama.2023.19296.
3. Thomas P. Quinn et al., "Trust and Medical AI: The Challenges We Face and the Expertise Needed to Overcome Them," *Journal of the American Medical Informatics Association* 28, no. 4 (April 2021): 890–94, https://doi.org/10.1093/jamia/ocaa268.
4. Sam Altman, "The Gentle Singularity," *Sam Altman's Blog*, June 2025, https://blog.samaltman.com/the-gentle-singularity.
5. Internet Matters, "Children's Experiences of Nude Deepfakes," October 2024, https://www.internetmatters.org/hub/research/children-experiences-nude-deepfakes-research/.
6. Internet Matters, "New Report—Estimates on AI-Generated Nude Deepfakes," Internet Matters, October 22, 2024, https://www.internetmatters.org/hub/press-release/new-report-uk-teenagers-encountered-ai-generated-nude-deepfakes/.
7. Jonathan Haidt and Eric Schmidt, "AI Is About to Make Social Media (Much) More Toxic," *The Atlantic*, May 5, 2023, https://www.theatlantic.com/technology/archive/2023/05/generative-ai-social-media-integration-dangers-disinformation-addiction/673940/.
8. Joshua Bote, "Snapchat's ChatGPT Bot Gets Blasted for Worrisome Exchange with Child," *SFGate*, March 15, 2023, https://www.sfgate.com/tech/article/snapchat-chatgpt-bot-race-to-recklessness-17841410.php.
9. "US Mother Says in Lawsuit That AI Chatbot Encouraged Son's Suicide," *Al Jazeera*, October 24, 2024, https://www.aljazeera.com/economy/2024/10/24/us-mother-says-in-lawsuit-that-ai-chatbot-encouraged-sons-suicide.

10. Jonathan Haidt and Eric Schmidt, "AI Is About to Make Social Media (Much) More Toxic," *The Atlantic*, May 5, 2023, https://www.theatlantic.com/technology/archive/2023/05/generative-ai-social-media-integration-dangers-disinformation-addiction/673940/.

11. Robert S. Mueller III, Report on the Investigation into Russian Interference in the 2016 Presidential Election, US Department of Justice, March 2019, https://www.justice.gov/archives/sco/file/1373816/download.

12. Daniel Mochon and Janet Schwartz, "The Confrontation Effect: When Users Engage More with Ideology-Inconsistent Content Online," *Organizational Behavior and Human Decision Processes*, October 7, 2024, https://doi.org/10.1016/j.obhdp.2024.104366.

13. Janna Anderson and Lee Rainie, "The Future of Human Agency," Pew Research Center, February 24, 2023, https://www.pewresearch.org/internet/2023/02/24/the-future-of-human-agency/.

Chapter 8

1. Sarah Min, "Google to Pay $170 Million for Violating Kids' Privacy on YouTube," CBS News, September 5, 2019, https://www.cbsnews.com/news/ftc-fines-google-170-million-for-violating-childrens-privacy-on-youtube/.

2. Rikki Schlott, "Meta Wouldn't Risk 1% Revenue Cut to Keep Teens from Harm," *New York Post*, November 25, 2023, https://nypost.com/2023/11/25/opinion-meta-wouldnt-risk-1-revenue-cut-to-keep-teens-from-harm/.

3. Julie Scolfo, "These 'Toys' Are Killing Our Kids," *The Hill*, March 2024, https://thehill.com/opinion/4438362-these-toys-are-killing-our-kids/.

4. BBC News, "Facebook to Face UK Fine over Data Breach," BBC News, October 24, 2017, https://www.bbc.com/news/world-us-canada-41488081.

5. Center for Responsive Politics, "National Rifle Assn," OpenSecrets, accessed April 30, 2025, https://www.opensecrets.org/orgs/national-rifle-assn/summary?id=d000000082.

Chapter 9

1. "Old Order Amish Mennonite Church," *Encyclopedia Britannica*, last modified March 30, 2022, https://www.britannica.com/topic/Old-Order-Amish-Mennonite-Church.
2. Courtney L. McCluney et al., "The Costs of Code-Switching," *Harvard Business Review*, November 15, 2019, https://hbr.org/2019/11/the-costs-of-codeswitching.
3. Cortney S. Warren, "How Honest Are People on Social Media?" *Psychology Today*, July 30, 2018, https://www.psychologytoday.com/intl/blog/naked-truth/201807/how-honest-are-people-social-media.

Chapter 10

1. Thania Siauw, "What Does Therapy Feel Like? 10 Insights to Consider," *Little Window*, September 17, 2024, https://www.littlewindow.com.au/blog/whatdoestherapyfeellike.
2. Savita Malhotra and Swapnajeet Sahoo, "Rebuilding the Brain with Psychotherapy," *Indian Journal of Psychiatry* 59, no. 4 (2017): 411–19, https://doi.org/10.4103/0019-5545.217299.
3. Phillippa Lally et al., "How Are Habits Formed: Modelling Habit Formation in the Real World," *European Journal of Social Psychology* 40, no. 6 (2010): 998–1009, https://doi.org/10.1002/ejsp.674.
4. Ashleigh E. Smith, Carol Maher, and Susan Hillier, "Here's What Happens in Your Brain When You're Trying to Make or Break a Habit," *The Conversation*, March 14, 2023, https://theconversation.com/heres-what-happens-in-your-brain-when-youre-trying-to-make-or-break-a-habit-201189.

Chapter 11

1. Amanda Reill, "A Simple Way to Make Better Decisions," *Harvard Business Review*, December 5, 2023, https://hbr.org/2023/12/a-simple-way-to-make-better-decisions.

Chapter 12

1. W. Island, "People of Color Face Tremendous Stereotypes During Times of Catastrophe," *HuffPost*, October 2, 2017, https://www.huffpost.com/entry/the-media-continues-to-stereotype-people-of-color_b_59d29ba1e4b043b4fb095b9a.

2. Tiffany Brockworth, "'Racist' JC Penny Ad Labels Black Family as 'Single Mom' and White Family as 2 Parent!!," *Media Take Out*, April 17, 2024, https://mediatakeout.com/racist-jc-penny-ad-labels-black-family-as-single-mom-and-white-family-as-2-parent/.

Chapter 13

1. Sharon Parrott et al., "House Republican Agendas and Project 2025 Would Increase Poverty and Hardship, Drive Up the Uninsured Rate, and Disinvest from People, Communities, and the Economy," Center on Budget and Policy Priorities, August 2024, https://www.cbpp.org/research/federal-budget/house-republican-agendas-and-project-2025-would-increase-poverty-and.

Chapter 14

1. Custom Market Insights, Global Self-Improvement Market 2024–2033 (Sandy, USA: Custom Market Insights, November 30, 2023), https://www.custommarketinsights.com/report/self-improvement-market/#:~:text=Reports%20Description,to%20reach%20USD%2090.5%20Billion.

Chapter 15

1. Abeba Birhane, "The Unseen Black Faces of AI Algorithms," *Nature*, October 19, 2022, https://www.nature.com/articles/d41586-022-03050-7.

Chapter 16

1. "Customer Acquisition Costs by Industry (2025)," Shopify, July 29, 2024, https://www.shopify.com/blog/customer-acquisition-cost-by-industry.

Chapter 17

1. "Democratize Tech: Mission-Driven Capitalism Is the Future," *Mother Jones*, October 2022, https://www.motherjones.com/media/2022/10/democratize-tech-mission-conscious-capitalism/.

Chapter 22

1. "12 States with Teens' Social Media Regulation—Is Yours One of Them?" Investopedia, December 2024, https://www.investopedia.com/states-with-social-media-regulation-for-teens-8757983.
2. "EU Artificial Intelligence Rules Will Ban 'Unacceptable' Use," BBC News, April 21, 2021, https://www.bbc.co.uk/news/technology-56830779.
3. Esau McCaulley, "It's Time for a Boycott," *New York Times Opinion*, March 23, 2025, https://www.nytimes.com/2025/03/23/opinion/target-black-boycott.html; Pam Danziger, "Are Target Boycotts Starting to Take Their Toll?" *Forbes*, March 4, 2025, https://www.forbes.com/sites/pamdanziger/2025/03/04/are-target-boycotts-starting-to-take-their-toll/.

ACKNOWLEDGMENTS

My deepest gratitude is due to Laurenne Sala, without whom this book would never have taken shape. She is the epitome of expertise—in storytelling, research, writing style, tone, editing, and being an absolutely phenomenal human being.

Thank you to the team at Radius Book Group for their partnership, adeptness, and skill in producing this book.

To my colleagues at Lotic, thank you for your perseverance, hard work, and collegiality. Even when the days seemed the darkest, you helped me see beams of light.

My wife and three children deserve raucous applause for putting up with me, not just in the writing of this book but in general. Thank you my dearest Eileen, Alaina Ryan, William V., and Calvin Thomas. Thank you to my parents, who instilled in me core values of integrity, service, and problem-solving.

"Consider that you might be wrong" is a guiding principle for me. It's a belief so strong, I have it as a tattoo on my left shoulder. This book is built on the foundation of over a decade of collaborative work with brilliant minds, yet I acknowledge it may not have all the answers. I hope this text moves you—perhaps you'll find inspiration in it, or perhaps you'll find points of contention. Either way, I encourage you to engage. If you agree, help bring these ideas to life. If you don't, let it motivate you to find another way to bridge the socioeconomic gap without creating further division. Whether you agree or disagree, we need as many perspectives as possible to participate if we are to create a more equitable future.

Finally, this book is dedicated to my recently departed Onyx, a Portuguese Water Dog who spent countless hours curled into me while I worked on my laptop. When he felt that it was time for me to sleep, he would "herd" me to bed by using his snout to push my computer closed and nuzzle me to move up the stairs. I miss you, Onyx Purple Papi Sox.

Finally, this book would not have been possible without the financial contributions from Algae Bloom Corp, a 501(c)(3) dedicated to objective, expert, and ethical assessment and explanation of the potential opportunities, risks, inequalities, and biases of technology design and use.

www.ingramcontent.com/pod-product-compliance
Lightning Source LLC
Chambersburg PA
CBHW010329030426
42337CB00025B/4871